GUARDIAN 1

THE
ANSWERS

Knowledge received and reported by:
Fred and Linda Foster

 Fred B. Foster Publications
Sacramento, California

Copyright ©1984 Fred Foster
Copyright ©1984 Linda Foster
Library of Congress

Printed in the United States of America

ISBN 0-9613762-0-1

This book is dedicated to the

UNITED STATES OF AMERICA

This book is dedicated to the UNITED STATES OF AMER-
ICA, because we have the freedom to receive and express this
knowledge. The Guardian for Our Universe indicates to us
that in the UNITED STATES OF AMERICA there is a very
important development. Freedom for the entity to experience,
freedom of speech, the integration and working together of
many nationalities, and the willingness to understand and
learn new ideas. We must never allow anyone to take these
ideas away from us.

This symbol was given to us by the Guardian For Our Universe, just before this book went to press.

"This is to be utilized for you, and for your book, and for the people of Earth. This will represent what is within your book — the knowledge of the beginning, of the protection, and of the universal understanding of what is happening and what has happened. Of the merging of one dimension into another, and the mingling of the two that produced the three, the going forward of this energy pattern. This is also for personal use for understanding of energies. Understanding of energies is not complete there yet."

Topics To Be Discussed

These topics and others will be discussed at different times and will be found in different places throughout the book; and you will find that more understanding is gained each time they are discussed.

The Guardian For Our Universe
What is our learning on Earth?
How we came to Earth
Why your original knowledges have been forgotten and blocked by you
Why these particular knowledges are not on Earth at this time
Why getting knowledge to Earth is so difficult
God, the true meaning
Why the door to Earth is closing
How and why this connection for knowledge was brought about
Why we have guides
Workings of the aura
Correct understanding of karma
What happens after death
The blending of knowledge and discipline to raise your vibrations
Do animals have souls?
Edgar Cayce, his knowledges, where they came from, and his time spent with Fred in Atlanteah
Our connection to UFO's and space people
How Pyramids were used
Statues of Easter Island
Should we clone people?
The new energy light source
Many other topics of interest

The Guardian For Our Universe has answered all questions directly and to the point so they are easily understood.

FORWARD

Within the covers of this book you will find the knowledge that is given is far beyond anything that has been given to Earth at this time. It is now known that new knowledges are needed to heighten our awareness and vibrations for what is needed on Earth, as the door into Earth is being closed.

The Guardian for Our Universe wishes to make us aware of this. He is doing something that has not been done for thousands of years, and that is to give knowledge to us directly.

In discussing the way in which to give knowledge, it has been decided the best way would be in the form of questions and answers. The questions and answers are given so this knowledge is brought forward to you in bits and pieces so you may grasp and understand what has happened to us since the beginning. The knowledge you will cover at this time is but a small portion of knowledge that will be brought to Earth within this life span as this is a time of transition.

In the time spent with the Guardian, our feeling is that there is only love and understanding for us here on Earth by those in the Universe, and much help can be given to us from the Universe if we will just expand our minds and accept new teachings.

We emphasize that this book should be read in sequence in order to be understood thoroughly and we also recommend the reading of this book more than once for the raising of your vibrations and for full grasp of the knowledge.

We will say no more at this time as the quality of the knowledge in this book will speak for itself.

Word Definitions

ANIMAL BODY — What the soul developments worked with to create the body that now is man.

ATLANTEAH — What Atlantis was called in the beginning.

CHANNELLING — When a person on the Earth allows his guide to take over his body for short periods of time, to talk to, or answer questions for, other people.

CONDITIONING — Adjusting the vibrations of the aura. This was given to us in the beginning to maintain this connection, and that we may receive new knowledge and understandings at this time.

DIMENSION — A specific area that is a specific vibration. One vibratory effect.

E-AH - (EAH) — At the end of a word means "from the beginning", or "of the first".

FRED — The one who always asks the questions. Also he was the Atlantean teacher.

GUARDIAN FOR OUR UNIVERSE — The one who always answers the questions. One who has been guiding our Earth since the beginning who has total knowledge of all. The Guardian will explain this in detail through this book.

GUIDES — Soul entities who are deceased and are now giving guidance back to Earth.

KARMA — Is the Universal Law that is used for balancing of the scales for the soul progression, to help make the vibratory frequencies correct.

LA MUREAH - THOUGHT FORM — The first La Mureah was where the soul developments first lived in this dimension and was of a high vibration.

LA MUREAH - PHYSICAL FORM — The second La Mureah that was developed after taking the physical body. They created a third-dimensional place to live which was of lower vibrations.

LIGHT SOURCE — This is a new energy source that is now being given to Earth.

OTHER SIDE — Where the soul goes after death which is a different vibratory frequency.

"OTHERS" — Always refers to space people. Soul developments that are similar to us who have been working with us and helping us since the beginning.

"THIS ONE" — When the Guardian says "This One" he is referring to Linda.

VIBRATIONS — Energy patterns of the body and the aura that are changed through the refining of the bodies and looking and accepting of the new knowledges.

VOICE TRACING — Tapes that were made.

WALK-IN — When an entity takes over an already existing body, usually an adult, to continue his learning experience, and the original entity leaves the body.

Chapter One

(Tape 11)

Linda and I have not said much about ourselves in the beginning as we feel the important thing is the knowledge that is coming to Earth. During the answering of the questions that I have asked, the Guardian's answers will make you more aware of where we fit into this.

As we are sure that the reader does understand, there are many thoughts, knowledges and ideas coming through from many areas. The area this knowledge comes through has not been open to us for many thousands of years. The reason, explained to us by the Guardian For Our Universe, is that our understanding was such that we would not be able to utilize this knowledge wisely, but that has changed. The knowledge in this book has not been given to Earth in the past. This is explained more thoroughly throughout the book by the Guardian For Our Universe. This is only the beginning of the knowledge and even at the time this book was being published, more advanced knowledge had already been received.

All knowledges and understanding in this book have been taken directly from conversations between Fred and the Guardian For Our Universe. **Questions and answers have not been restructured because restructuring a sentence could change the understanding of the knowledge the Guardian is giving Earth.** We are sure that the reader is well aware of how this can happen. We have an understanding with the Guardian For Our Universe that **knowledge will be given exactly as we received it.**

The questions asked in this first chapter are mainly more of a personal nature, but were included as they give knowledge in areas we felt would be of interest to all. They are often discussed in more detail further in the book.

We wish to tell you what the Guardian For Our Universe told us: **"Listen and do not project your Earthly thoughts into what I am saying."**

All questions will be asked by Fred and will appear in this book in *italic* print.

All answers will be given by the Guardian For Our Universe in regular print.

The Dialogue

Are we allowed to ask questions about things in the Universe?

Questions are allowed.

We have had people on this Earth that have left a lot of memories and we have wondered if the memories are right or if people have distorted them? One of these people is Edgar Cayce. We have read in numerous books about him and he has predicted that by the year 2000 our Earth is going to have a very big upheaval. If this is going to happen we would like to help people to prepare for this. We would like to see that some things are preserved so that all is not lost.

You are free to prepare for what you please. To enlighten or to give knowledge is always of value.

Can you answer about Mr. Cayce?

His knowledge was given as it was seen at that time.

Has that knowledge changed?

All things change and do change. This may be expected as given, however, there may be some slight variances.

May I ask what the variances may be so that I may understand them?

This depends as yet on other things that may pass.

Does that have to do with the Earth itself? We understand that it is an entity, and that it is going through cycles. Or does it have to do with the consciousness of the people?

It has much to do with both and still other factors that are not understood by you. All things are not known in your plane of existence for understanding and, therefore, explanation may be difficult.

I wish to ask a question that a lot of people have been very curious about. We had a man on this Earth by the name of Jesus. Are all teachings and writings that are here now true or distorted? Was he a master teacher or was he the Son of God.

The man known as Jesus was indeed a Master, and still is.

Is he on this Earth at this time?

He is not.

Will he ever return to us?

At this time there are no plans of Jesus returning to Earth.

Are there any religions we should follow on this Earth?

Follow the dictates of the heart. Religion is a matter of conscience, of heart and of finding one's way.

We are taught about religion from when we are very young. By blocking our past lives so we may never understand who we are, and by keeping our minds closed to other teachings, how are we going to learn about the correct understanding?

There still is understanding of what one does. Understanding goes far beyond what you know on Earth— far beyond.

The understanding that is given to you is the understanding that is better known at other times. Things may then be looked on in a different way. Progress is always made in some way. Even though some may tell you that progress in a lifetime may go backwards, or not be made at all, this is not entirely true. Lifetimes always give some type of understanding that is known to the total consciousness even though it may not be known to the individual conscious mind.

We would like to know about karma, and how we progress from one lifetime to another, and in what manner we may achieve enlightenment.

Karma is the Universal Law that is used for balancing of the scales. To balance each soul, as the Egyptians once thought when one was dead, to see where that person was in that lifetime, is of importance. Balance and harmony throughout the Universe are of great importance in all things. The soul is not different. What is learned and what is known is to be shared and progressed on each level.

Each thing that is given is given for a purpose. This purpose can be carried out in any manner due to free will. Then each thing must be weighed and balanced to see if this has been handled correctly for each individual soul, which may vary from soul to soul.

Each lesson has been learned to some degree in each lifetime. One may require a good deal of knowledge and understanding in one area. Another soul may have required understanding that even may not be seen during the lifetime, and therefore only requires the understanding of only a portion of this same knowledge. It is very difficult, when on the Earth plane, to see or have understanding if a person is learning or has learned a lesson. This must be done by the individual soul at a time when it is free and unencumbered by the physical body.

Is there any way for me to progress myself further on the karmic wheel?

Much is misunderstood. All are now seeking to remove themselves from the plane that they are now on, but this is misunderstood. To remove yourself from the plane you are now on, while indeed there will be progression of sorts, into another time and dimension only means that you will be learning elsewhere other things in other ways. At this point in learning much understanding is being received and given by yourself, even. This understanding is easily understood and easily assimilated for your knowledge and progression, however, sometimes on other planes things are much more difficult. Therefore, people who are constantly looking to leave the Earth may often be seeking that which will be more difficult than it is now. All parts of learning are just that. They are all parts of learning and are to be considered a privilege to have these chances to continue to learn at this time. All knowledge is good.

Different understandings are needed by different people. All people have their understanding that is on the plane, or level, that they are seeking at that time. While unconsciously, or perhaps I should say "When they are free from their Earthly body," all may seek a different type of understanding. On the Earth plane these understandings and thought forms are put more into practice and are only tools of expression. Therefore, each person is looking and seeking for the expression that can best, at that time, fulfill himself.

If they are not able to use this higher knowledge, they are still using the highest knowledge for the progression of themselves at that time. So do not be that concerned if a person refuses a higher knowledge. They are at the highest knowledge that their energies can assimilate and absorb at that time.

I am finding a lot of young people are seeking more knowledge but they are not able to find it because, while there are good teachers on this Earth, they are not that numerous. Maybe this should not be a concern, but Linda and I would enjoy helping some of these people.

In their seeking they are releasing and if what they are finding is not what they are looking for, they will continue to seek. All knowledge that is given is good. If you wish to give this knowledge you may do so. There are those who will be glad to listen. There are always those that are on the same higher vibratory frequency as yourself.

21

Am I allowed to take knowledge from one lifetime to the next lifetime?

All take knowledge from one lifetime to the next. Without it you would not be where you are today.

We have learned about traveling our past lives. Is this the manner in which to bring it forward? Is there something that we can do in this lifetime so this is open to us in our next lifetime?

Prepare yourself at this time. When you leave your body and go to a different vibratory level (when you die) if this is still your will and desire and is still important to you at that time, on that plane, then you will seek the way to find knowledge when you come back again. You will prepare yourself, you will prepare your body and mind at that time to receive that information.

Can you give me a date of when Atlantis will rise and Earth changes will come?

With the changing of your century and the changing of your time period of which you have been given instructions. There will be much that will be found then. Not all at that time, but much.

Does this mean that part of Atlantis is going to rise but other parts of the Earth are going to sink?

Yes.

Is this to happen by about the year 2000?

Much will happen around that time, but more will come later even after the year 2000. This is not a thing that will happen at one time.

Will it be very much as Atlantis disappeared, little by little?

Parts, yes.

We have been very interested in crystals. May we have the means to go to where the knowledge of the crystals is kept? The teachers that you have already given us have been very good and we would like to know more about the crystals.

Crystals are living things. They are not objects as you think of objects. They are living beings evolving in their own time and space. These crystals in the time of Atlantis, and also other times, were found to be able to be charged to do certain things. This was within their growth period at that time. At this time the knowledge to find the ways to activate these crystals is not yet within the scope of mankind. These crystals will help you in many ways and are very much attuned to man. However, some things are not obtainable that were obtainable in Atlantis as different paths have been taken at this time. Different paths of knowledge and of knowledge of energies.

Are we not allowed to change these paths in any way?

The technology is not available. As understanding of the Sun's powers comes back into focus, then many things will be opened up to your planet; but until more studies and more advancements are made along that line, some possibilities in the crystals, of the many varied functions, are not possible.

Am I to be part of this at this time?

Not at this time. There is not sufficient technology in this lifetime. This technology has not been truly begun there.

Why has this knowledge of the crystals, that we had in Atlantis, been taken away from us?

This has not been taken away. This has simply been lost, as many things have been lost during the time periods.

Can any knowledge be given of the crystal that is not of the magnitude that would change things? Such as meditating on the crystal or putting your own energies into it?

Much knowledge of the crystal for small purposes is known. There is not much to be said different than what I have said until the path is prepared through more energy technology from the Sun. There is no way to give you more understanding of the inner-workings of the crystal. The crystal is only a symbol of clarity and purity. This is only a symbol to meditate on. But there are certain energies from the crystals that do help make your thought patterns more pure and more accessible and concentrated for these thought patterns to form. This must be understood also that this is a natural crystal that we are talking about.

Are there any crystals coming from certain parts of the world that are better crystals for what we are talking about?

This you will have to attune yourself to. This is because of different frequencies and vibratory patterns from your own being, not because crystals from one part of the world are better than crystals from any other part of the world.

Does it have to do with where you have lived in the past or in past lives? Your energies are in tune with crystal from that place?

Past energies are incorporated in energies of today. That makes **you.** Of course that is part of it, but again, this does not mean that because you were from Atlantis at one time that Atlantean crystal would be better for you. You have been from many places and many energies are now combined and many thought patterns are combined and emotional wavelengths have been combined. These change, these continue to change and develop with each lifetime so each lifetime and each day even they change somewhat. So, when you get a crystal and you work with it, your energies do become in tune with this crystal. Then, when people talk of a crystal being "their" crystal, this is why. The energies become attuned between you and the crystal. This is what makes a crystal powerful and to work for a person, as it is with any gemstone. Your vibrations and its vibrations combine to make it yours and to make it work for you.

Does the size of the crystal have anything to do with it?

That depends on what you would need to use it for. Sometimes size would be of importance but do not work with a very tiny crystal. You need something that will put out energies and, of course, a very tiny crystal will have very tiny energies. Although, it isn't necessary to have a huge one either. One that feels right with you.

Before La Mureah and Atlantis, were there other worlds that had reached the magnitude of knowledge that they had?

This planet that you are on is young in comparison to many others in your universe. While La Mureah was the earliest advanced civilization on your planet, there have been others at that time but were very small and did not continue or it was a very long time before they amounted to much. Therefore, knowledge that would be remembered, or found, would be from another place. La Mureah, as it was called, although it may be hard to understand, was of a different technological era than you understand today and also different from Atlantis, although some of Atlantis' technology did come from La Mureah. Much of what was in La Mureah and Atlantis only remains as thought and memories that were not carried on. Many things that are to be found later, through the Earth's periods, will be back to the technology of that time. As you have advanced in the civilization in one area, in one way, there are many ways yet that you may go; and going back to nature, as it is now termed, is the essential way to raise a civilization to the heights that it had been before, and still not continue to contaminate yourselves as beings on your planet.

There has always been intelligence on your planet. The time of La Mureah was of very, very, high intelligence, because they were originally the beginning of and the connection in with the creative source, of sorts, as you understand it. Therefore, their knowledges, their feelings, their prosperities were different. They were more in tune; and when people talk of sin or falling away from God's grace, it is because this has been lost. The thought, the technology and also because the feeling and the understanding is gone.

When you continue on the search that has begun recently of going back to nature, going back to once what was, then as you continue to go down these paths you will begin to look for new avenues of understandings, and the further you get into this the more understanding there will be in these other ways. Going back to the creator, or the God-head, is not so much as going back to a place but going back to a consciousness. This consciousness, at that time, will then give you the things you have been seeking and looking for. Remembrances of past lives give you the feelings and emotions that you have separated from something. Looking for these things outside yourself, as has been done for many, many centuries, is not the way you will get back to the creative source.

Now things are beginning to evolve in this other direction. Continue to look within. This is one of the greatest ways that you will begin to understand knowledge. To look within has always been known through the centuries but has been kept secret or has not been passed on, as many did not have understanding. As knowledge continues to rise, people are beginning to realize that through many lifetimes the search for this has become futile, and are now seeking new ways to find an approach to be able to try to go back so that they may be one with the source so that their life may continue on and become what it was in the beginning and what it was meant to be. This is most of the understanding.

It is good at this time that people are looking for love, understanding and to learn to harmonize with everything that is within nature. It has been said that God has given you everything. You do have everything. It is not that you have lost it, but that you have lost the understanding and the ability to find it.

We are sure that the knowledge that you are giving in this lifetime will help to change this.

But this is not to be done overnight. This took a very long time to get away from. It is taking a long time to find the way back. And this is not to be found outside of yourself. This is only to be found within yourself through love and compassion and understanding. Many have come and many have shared their time there to teach this knowledge and it is not understood yet! Therefore, the looking within and the changing of the thoughts, attitudes, love and understanding, these are the things that must be done individually and will be then taken collectively to make the changes necessary. Every individual one person does make the difference. Every one individual thought form and every individual learning process of loving and learning will be collected together, and as these thoughts, patterns and emotions are changed through lifetimes, the collective energies will make the things you wish possible. However, these things do take time. Things do take understanding and people must change individually so that there will be a collective change in your universe.

Understand that these are the inner-teachings of the Christ that have been missed but are now being brought forward by many people who have been changing their thoughts and attitudes for several lifetimes. This is not a thing that has just come into being all of a sudden. This knowledge has gradually increased, has gradually been brought about until, at this time, there is a large enough consciousness that it is being spread faster and over a larger area, and people are beginning to understand more quickly.

As this consciousness grows and as people put it into practice in their everyday life, the attunement of themselves as individuals and collectively, then what happens over your Earth is what makes the changes and what makes things happen correctly in your world. It is not only external. External is but a manifestation of the internal.

NOTE TO THE READER: A reminder that this book is a conversation between Fred and the Guardian For Our Universe, and that sentences have not

been restructured so that the knowledges can be understood as given.

*We appreciate your taking time to talk to us and give us this
information.*

This is well within what I choose to do.

*I'm wondering, if before I was on this Earth, if I had duties, a
job? What was my being? I don't feel that when I came to Earth it was
my beginning. If I understood where I had come from, I will partly
understand what I am going back to.*

What you have come from is as varied as what you are
now. Being free as a soul before you came to Earth, there
were many things that you were capable of and many duties.
However, the time just before coming into the Earth's vibrations was a period of freedom and rest and therefore no
duties, as you say, were required.

*When you say duties, I get excited because I feel there are a lot of
things out there that I myself would like to get back to. This feeling is
very strong within me.*

This is very curious, all seem to want to go back, and yet
all wanted to come.

*I'm not saying that I particularly want to go back. I'm saying that
if the Earth is going to be my home from now on, then I should work
towards making this a better place for me to live this lifetime and my
lifetimes to come.*

That should be the object of everyone's thoughts.

*Is it that we should not be wanting off the Earth, to go back to
where we came from?*

It is not your primary goal at this time to be "off your
Earth." Lessons are to be learned in this time period. Accept
what you have and make the most of it. There are other
dimensions and other planes of existence that are not familiar

to you now, but will be familiar to you again; however, this has been chosen at this time, by you. Your learning, your progression, everything that is for you, is there now. Do your best and learn all you can as quickly as you can, that is what you are there for. This is like wanting to go to school, and before kindergarten is out you are anxious to go to college. This is a time period for learning and for growing. And when you have learned and grown all that you can at this time, you will then be summoned somewhere else for other duties and other learnings. This is a time for your growth and not a time for rest.

I was just curious about what my life was before this Earth.

Your life was bound on other places at different times, learning different things, experiencing things. Life away from your Earth is in many ways similar to life on your Earth. There are still times of play, times of rest and times of learning. However, learning experiences are totally different. This is why it is good to learn each as you have the opportunity. It is like going somewhere and paying to get in and anticipating what you will find out and before you learn or before you have found out what you have paid to see, you leave.

No, I don't want to leave, I'm learning and I want my conditioning to continue on.

Conditioning will continue and very soon you will see it will be accelerated.

I feel that it is being accelerated now, changes in my diet were very easy for me. Easier than I thought they would be. I've been eating meat all my life and I have found it very easy to quit. I feel the conditioning has a lot to do with it.

I am happy we are able to help you.

We are very curious people on this Earth and there is knowledge that we do not have and maybe should not have.

All knowledge is for having; however, it is necessary that you understand the knowledge that is given or it is of no value to you.

We have always called you My Lord. Do you have another title in the Universe?

In the Universe, titles are not as important as titles on your Earth plane. My Lord is significant and will be fine as "This One" (Linda) has called me that for a very long period and is well acquainted and the vibratory frequency at that point is quite acceptable.

> NOTE TO THE READER: At this time I would like to explain his statement further. As spending many hours with My Lord, I understand that titles are not important to them, that there is love and understanding in the Universe and everyone does his job. In talking with the teachers he gave us from his Universe, they refer to him as "The Guardian For Our Universes" and even further in the book he does refer to himself as being the Guardian For Our Universe. As we have worked with him in other lifetimes titles were never mentioned. We have received knowledge and help from him before, but not in the way we receive knowledge from him now, as this is now coming direct. As more knowledge is given in this book, on these subjects, you will understand more about the way of the Universes.

Are there other people on Earth who come to you?

On this Earth time and period "This One" is the only one who is coming to me. She has been here before and we have worked together long before this lifetime.

Have you ever worked with me before?

Yes, but this is not the same.

I wonder if there is any way that we can know how we are doing in our progression?

When this time of existence is over you will know how you have done. You know now, this is not a secret. You understand what is going on. Progress is felt and understood.

We have been trying to analyze our dreams. We have understood that dreams are the working out of our problems. That our body and mind work out a lot of problems, and we are trying to understand our dreams more, so we can understand ourselves more. We have talked to different people on this plane about our dreams, but we seem to be getting different answers from different people.

Dreams are a personal unit. However, different interpretations are for different people. Basically this is communication. This is problem solving. But it is much more also, and while trying to interpret your dreams is very important and can help you in many ways, this is not fully understood at this time on your plane of existence. To interpret your dreams as you see and understand is wise and, for health reasons especially, can be of much help to a person. To understand your progress as you are going, it is important to listen to yourself. Also, as your conditioning progresses I will tell you more details and you will understand more.

Would you prefer to give us a teacher to help us with our dreams at this time?

Of course there are teachers in this Universe that I may give to you.

May we have the way to go to a teacher who may help us with this?

You may go to Bhalor.

May we have the way to go to Bhalor, the symbols we will use to go to Bhalor?

> NOTE TO THE READER: Symbols cannot be given here because if the mind is not of proper frequency it could cause trouble. As knowledge of dreams is made available to us, we will make it available to the public.

Is there a time we should come to you that is better for you?

It would be better to set a time and to meet that time, therefore I may continue on as I wish. It would be better to come when "This One" is not tired. There has been no limit or time limit put on questions.

Yes, we appreciate that. You have often mentioned Earth time; is your time so much different from ours?

In the Universe, as in your world, there is time. There is a time for things. This is not a place where there is no time. Some kind of time is everywhere. However, do understand that your time and my time is not the same. Your time is very tiny compared to my time. Many thousands and millions of your years go by in a very short period of time for me, and this is true for you also, only you are ruled and regulated by the events and the time period you are now in.

After we go to the time of truth (when we die), there seems to be a variation of the time interval when people return. Does that have to do with our birth rate here? Or does that have to do with some people wanting to return faster than others? Or is there something that governs our returning to Earth?

Very much governs this — your desires for one; it is an important thing. Once you have made transition — your time to realize and work through problems, to help you better understand, and to allow you time to understand what you have done with that lifetime, and what progression you have made — many are anxious to go back to make corrections in this area, as these corrections are more quickly worked out when you are reliving and reworking on Earth. Time to have realization of what is going on, on the other side, may take longer. The resting period may be longer — there is no set pattern. This is determined much by your feelings.

Linda and I always felt that we were together for a reason.

This is true, and for many reasons of Earthly ties, not of ties before your Earth.

That helps us understand more. We would like to help teach the new understandings to others. Some seem to be searching but are not able to find the guidance.

Teaching is of importance to the one who is being taught. The degree of that information or importance is put on by the one that is being taught; what is needed for them. While it may be very simple to the teacher, it may be of great importance to that one. There is no way to put value of more importance on teaching. This is individualized. This is for each one to decide. Many times a very simple statement, or a very simple thing to you, will be of great importance to another. Higher teachings, as you say, are for people who have already learned much and would be ready for those types of teachings, and there are people who are ready and are not receiving those teachings. But do not think that higher teaching to them is any more important than it would be to someone with lesser knowledge — someone that would receive simple knowledge.

I am finding, in talking to people, that many do not understand.

That is true, but do understand that all must start somewhere. Do not bother those who do not wish to listen, this is of no importance. If they choose not to listen it does not matter. And understand this also, as you are able to understand more, I will give you higher teachings and more knowledge. You cannot step from the beginning to the end, there must be the middle and you must gain the middle ground to get to the end. So, teachings from the very beginning to the very end are all very important to different people and all must be given. It is not for you to judge who is ready for what. This is their own inner-workings that will make that decision. Teach to all — you may wish to have teachings for different levels, but do not rule out those who are beginning to learn. Present to all. It will be accepted, each in his own way, or rejected in his own way, and all is fine.

In presenting these teachings to our children, we are finding some wish to go on their own path, what about this?

These children that we discuss now, understand, are children now only. They have been brought together for a reason, and guidance is being given that each one needs. Each one has their own particular path. They all have had much knowledge and much understanding in the past, and also much karmic dept to pay, and much karma to work out between themselves, as it is in all families. These children have also been given the chance to continue, and the chance of knowledge that many will not have. They chose this in the beginning before this lifetime, so that this time they may continue with their studies and their learning of progression so this lifetime would not be wasted, and would not be of learning Earthly lessons only. These things are done with purpose. Then if they are ignored, through free-will, it becomes even more difficult later, because the chances that were given, or were set up to be taken, were not fulfilled. You must always do what it is for you to do or it becomes more difficult the next time.

Do not think that what is done on Earth is different than what is done everywhere else. Things are much the same. All have progression. All have things to do, and all have things to do at various levels. What we choose to do is in a state other than the conscious. This is on every plane, every dimension. All must do the best they can, with what they have so they may have more to do with, more to learn with.

I understand that there are actually other planets such as Earth in the Universe, and eventually, as our understandings are raised, we are able to go and learn the teachings in that particular world.

There are places throughout your universe. These universes were made for you and for me. Your universe was not made and then by chance you were put to inhabit it. This was made for a purpose, and you were that purpose. There is every type of learning, every type of play. The different things and different places are innumerable. You spoke of higher learnings. I understand that this is so and this is true for you. True, some learnings are higher, but do not think that where you are is of low understanding or of low learning. Much is learned on your plane. Much is understood there and in other places much is understood and learned also. This is a continuation of doing and of being. This is not as you think of as more, better, lesser. This is continuation of all learning, of all understanding. As we discussed earlier about people with less understanding, and in teaching people of higher understanding. It is all learning, and all is understanding.

We want to thank you for making yourself known to us in this lifetime, so we may ask these questions.

This lifetime or other lifetimes, it is well for me to be known to you, and will continue to be known through "This One."

Then there is no chance for me to come to you?

There is no reason why you could not. You and I have worked together several times through her before, and vibratory patterns and frequencies are now in tune. This pattern is now set up and at a later date will be of great importance. This will be at a time when you will not be with her, and will be in need. The pattern is being made for future times. Therefore, you see, each thing that is being done is for a purpose.

Is there any other way of going to you, other than the way I help her to come to you?

Many, many years ago, many hundreds of thousands of years ago, when the time came that "This One" was needed to be in contact with me again, I brought her essence to me and told her that when I sent her back, to go to her Atlantean teacher and tell him what happened, he would give her a way to change her frequencies to be able to come to me in another manner. The way that she comes now, this is not a thing to say "I wish to talk to you." No, I do not converse with just everyone as vibratory patterns must be correct, however, as you are being conditioned you will have your own way. You may use the same way that I have given her, because this has been made so, but you may find your own way later if this is not satisfactory for you. This has been ordained and this has been set up as being so. Time will come when it is necessary. (Fred was the Atlantean teacher.)

Does it help for me to open my mind before I go to sleep? Is some of the conditioning that you are helping me with done at night?

It will help.

Does some of the conditioning that is going on have to do with my dreams?

The conditioning will affect your dreams, yes.

Does conditioning of this magnitude take more than one lifetime?

This depends on many factors. It depends on the magnitude of the conditioning, of what is desired to be done, your willingness to cooperate, the conditioning of the whole vibratory effect of the body, the mental and emotional. All these bodies must be in tune. Not only the thought of wanting this to happen, and working at it so that it will happen, but you must also attune each of your bodies and each of your vibratory patterns and frequencies so that they will balance and harmonize. You cannot allow part of yourself to go one way and part another. This is a fine tuning of the whole mechanism.

Is there anything else that you can tell me towards helping me on meditation that I may do myself?

It is better, when you make a decision to meditate, to continue on with it. Always continue on. Do not say "This was not a good meditation" or "This was a good meditation". You do not know if this was good or bad, for sure. You are looking on one level and what your mind is seeing is not necessarily what is being done.

Each day the meditaiton should be continued. Many say karma is released through meditation, but that is not exactly correct. Much tuning of the body is done through meditation. Much of it is bringing the vibratory patterns of your different bodies into a more perfect vibration. The doing of good or evil is another thing.

Karma, as everthing else, is on different levels. Not only good or evil, but it also has to do with the vibratory rates and vibratory patterns, as everything in your world is vibratory rates and vibratory patterns. Therefore, harmony and peace must be brought about into the body. That is much of the reason for meditation. When these things are fine tuned and you are more peaceful, your mind is more peaceful, then your body is also more in harmony. All these things will work together to release much negative karma that has been built up through being out of balance and out of tune. Relaxation, contemplation, and meditation — these things are of great importance. The physical effects it brings upon a person outwardly, are only the things people may see, and may prove. The effects that are upon the bodies that you cannot see and prove are the most important. Continue on with your meditation.

When my conditioning is over will I just realize it's over, that I've learned what I need to know?

Proper conditioning works like proper meditation, it continues on and continues to build a finer and a more perfect species. It all works together. I would say when you know that your conditioning is over you will be at a point where you will not need to go back into that vibratory pattern again. This may take many lifetimes. Conditioning that has been requested will be continued as long as it is desired. This is not a thing that begins and ends within a short period, and all of a sudden, as some magic trick has been played, you come about and become the thing that you wish — no, that is not correct. Knowledge of wishing and wanting and asking for this conditioning, in a lifetime, is of great benefit to you. Not only do you realize what is happening, not only do you request what is happening, but all these things will accelerate what is taking place. You see when you are not aware of what is happening to you, it is a very slow process. When you are aware that this is what you want and consciously work on that level, then everything is accelerated. Then the knowledge of desiring something and realizing it is brought into play. Then this all works together and becomes the whole.

Do not think that only because life after life is spent and some progress is always made, as stated before, that this means that progress is being accelerated. No, it does not necessarily mean that. Knowledge of what is happening and of what you are doing does accelerate this as many disciplines of different kinds accelerate any type of advancement that is being made.

Does it help me to work with other people or should I just be concerned with myself?

This is a very general answer. Of course the most important thing that one person has to do is to have full knowledge and full understanding of himself. But to help others is also of great importance, if they request it or if they desire it. You cannot force your thoughts and your beliefs on another. However, it is good for others to know of this because, although they may reject the thoughts and the knowledge at this time, that is their choice. They may do this if they wish. They may hear this many times before this begins to be understood by them, and they begin to have a better understanding within themselves, but each time it will be understood better. Also, it is good to discuss matters among those who do think as you do, so you may all share your knowledges.

When bringing this knowledge forward in book form, I would like to use the 7 dimensional teachers that you have given us along with yourself. Is that alright?

This can only be done with the full knowledge and the full agreement of each teacher.

NOTE TO THE READER: All knowledge in this book has come directly from the Guardian For Our Universe.

If I should need a teacher for a certain book, may I come to you and request this, as I have several books in mind?

This will be discussed when you are ready.

There are a lot of discussions here on Earth about UFO's. Are they coming from other dimensions to see about our well being?

Many worlds, and many galaxies and interplanetary sources are alive and within the scope of your dimension.

Then some of these spaceships are coming from other dimensions — from other galaxies?

These are both possible.

Is it possible for a person to make contact?

Many people have made contact. Many Earth people have seen and talked and have made contact with these.

I would very much like to make contact with them.

That is good; however, do not lose sight of the fact that you do live on your Earth, and you do live in the time plane that you are in now, and do have a life to live. Much progress will be lost if you continue on a line where you neglect to continue on as you have chosen.

No, I don't intend to change that. I really thought meeting the space people would increase my knowledge.

You must understand that this is something that must be done by you, when you learn to make your vibratory patterns in a more one pointed way. Also, when you can get more in contact with yourself through mental imagery and mental thought. This for you is a possibility, if you wish. If this is your desire, this then is your task.

I have a lot of ideas as my mind wanders in many avenues. As my vibrations are becoming more blended will I, more or less, lose all these ideas?

Not at all. The ideas you have that continue to run through your mind consume much energy, waste much energy and many of these ideas are not workable. When you learn to control and to make youself more harmonious, many things that would have taken up much of your time and energy will be discarded immediately and will not continue to waste your time. Therefore, much thought and much time may be given to those things that are obtainable or those things that you do wish to pursue. This makes your life much more productive with much less effort, therefore makes things flow much better and the ideas that you do have will be of a superior nature.

We are, at this time, on a meat-free diet called vegetarian. I feel this should help us in some of the things we are doing. Is the not eating of meat an important thing on this plane?

You must understand that all things on your level do have vibratory elements that are being put out. Eating flesh and eating meat is of negative vibrations. Things are given to you to eat that are of a different frequency and of a different vibratory pattern, and there is much difference between fruits and vegetables and animals. Also, you are within the animal vibratory frequency where you are encaptured at this time. Eating of this flesh is not recommended. There are instances when the eating of flesh would be necessary for the continuance of more flesh and there could be some periods and some instances where this would be acceptable to continue in the life force, however, much knowledge has already been given, that you are aware of, concerning this. You realize that eating flesh at this time is not good. To continue, once this realization is totally understood, is not good.

Do continue on the path you have chosen. You will find that your vibratory frequencies will also be aided in this. Those who do not eat flesh are more passive people, as a rule, and eating of flesh is said to make one more agressive and this is true, but more than just externally it does change vibratory patterns and frequencies and many other things. I would say that for you in this time period, that not eating flesh would be of great importance.

Are there any vegetables or any foods that are more helpful or have proper vibratory rates that are more harmonious for us?

With the makeup of the body, rice is very good for all. But each person is an individual and must utilize what is necessary for that particular body. But all fruits and vegetables are allowed.

What about herbs and vitamins that people say help the body?

And this is true, also the needs of the body change from time to time. There are tests on your level to help you determine these things, if you wish to know for each individual.

Is life with our family very important to our vibratory rates?

Life with everyone you are in contact with is of importance, of course. With your family it is of much importance because you went there to learn and to have better understanding with them. Not only of souls and people, but of situations that are at hand. Negative situations are often repeated over and over many times until you learn to handle it. Until you understand it. Not until the situation changes **but until you change.** Then that situation will be a simple matter for you to handle.

I believe "This One" is getting tired. Thank you for the knowledge that you have given us today, and go well.

Chapter Two

(Tape 12)

In talking to you about our other teachers that you have given to us, you said that we should receive their permission before we do certain things. We have played some of our tapes for other people who are interested in more knowledge than they have at this time. Does this meet with your approval?

At this time this does meet with my approval.

Thank you. Can you tell us a little about La Mureah, how the understanding then differs from the understanding at this time? We would like to get as much information in this area as possible before we move on to something else.

La-Mur-e-ah, At-lant-e-ah, these names and others that are similar have been brought forward from a very early period. E-ah is symbolic of a statement that is made "from the one", "from the beginning", "of the original source", "part of the beginning". These are terminologies, as can be said, for you, so that you can understand. So a word with that at the end does mean that this place is of the beginning, from the original source. La Mureah had a beginning before its name was conceived or brought into being. Do you wish that also?

Yes, we wish that also if you would.

As the world, your Earth, was forming and spinning very fast, the dimensions that surround the dimensions of Earth were found that, in one certain portion of the Earth, as the vibrations were changing and being created and being rearranged, one portion of that area of Earth was where the other dimension could be part of your dimension, or as a window to see through from one to the other.

This made many curious, and to be able to see into the other dimension was raising much curiosity. Therefore, many began to contemplate the other dimension also. As curiosity became more and more, some of the souls began to go through the portal and found that this was handled very easily. Dimensions of vibratory effects were not as great as had been anticipated, therefore, short periods of looking and enjoying were observed.

This opened up a whole new existence and experience. Many new thoughts came into being and, as one stayed longer and longer, they began to see that they could stay on the other side and were able to continue their patterns, although in a different vibratory pattern, but still continue comfortably. Into this new world they brought forth the thought forms and thought patterns of the other dimension and were able to set up a world of thought forms there. As this continued on further, it was continuing to be harder to go back into the higher vibration and, therefore, became easier to stay there longer.

This created many more new avenues of thought and development. New challenges were seen, new things were being expressed. Observing the different types of vegetation and animal lives that were there was quite fascinating.

As this progressed, some decided to try to lower vibrations even further to expand their experience there. There were many life forms that had already begun. The changing, and watching and then experiencing for very short periods, into these life forms, was amusing. While the world had been created there, they were now exploring into this dimension much further. Many found that this existence also had other soul developments and as they were being able to see, understand and contact these other soul developments, there was some dissension between the two.

The souls wanted to experience more and realized that the vibrations they had were not sufficient to have the experience that they desired. Therefore, there were some who began to make adjustments into these life forms. While controversy came, there was eventually an agreement that was made, and the one life form that is known, that is now housing man, was agreed on. The agreement was made that this body would not be taken by another soul development. That this would be free for the souls to come to experience in this area, as all others had also. This was a give and take experience, and this way all could be satisfied. Therefore, changing of vibrations of themselves and the creatures that they had decided to take over, was not an easy process, as they thought it would be. As this was being done, it was being done mostly on the thought level. But some that experienced this were in much more of a hurry and tried to experience it before they were ready. When this happened there were some souls who were damaged. These souls who were damaged were then returned to the source for re-developing themselves.

As this new housing that they were trying to find was now becoming a necessity, the vibratory effects were not as they were, and they realized they must find a housing as they were not able to continue in this lower vibration; and to get into a higher vibration again was becoming more and more difficult. So the taking of these bodies was as much out of necessity as out of curiosity. Through this practicing and working, they finally realized that the vibrations must be lowered to continue their existence in this dimension. By this time much advancement had been made, but also there were many that could not make the change. They lost much of the connection that had been before with higher thoughts and higher ideals, as you call them, because they were of higher frequencies and vibrations; and, when encasing themselves within these bodies, it was not as satisfactory as it had been desired. With some the change did take place and was a success. At this point, you were having several different degrees of success and several different degrees of not success. As the ones who were successful began to quickly realize, there was establishment of the higher and now of the lower vibrations and much of what they did in this land, at that time, was begun with the realizations and higher understandings of both.

Those who were not successful did not usually stay in this same area. They were placed in other areas to continue on through their advancement as it was being done. At the time that the world was still slowing down, and was still changing, the portal was eventually shut off, as it were, and those who were there, while still with remembrances of what they could do, and still could do, began to make their home in a land they called then La-Mur-e-ah.

There were many various teachings and learnings. These were continued to keep the high ideals and the thoughts of what had been and what they then wanted to try to regain, as they had not intended to lose what they had to gain this. This was not the original intent. As the new La Mureah was built in the fashion of your dimension, still much was able to be kept of their knowledge.

While those who went out had lost this knowledge, mostly to a degree that there was no knowledge, but there were the intelligences and changes that had been made, and began to learn and to experiment gradually. These were in different areas. They developed each in their own area and each with their own ways — a survival of the best ways for those times. As this was all very young and all very new, many mistakes were made on all levels.

At this time it was decided that there must be guardians and those who would help to develop or to watch over and to see how this experiment, as it had turned out to be, was going to be, as this was not a planned, definite arrangement as normally is considered in the Universe. During this period, the ones who were to watch over this, and to guide, were then put in charge and this was begun. "This One" here, (all through this book when he refers to "This One", he is referring to Linda), was very capable and was watching and controlling some of the mutations that had been developed.

The entities, while very young and trying to learn between the two dimensions, were very vulnerable in many instances, as were the ones who had no remembrances. Therefore, a watch was made to help them as they were being quickly overrun or destroyed in these bodies, as it were. Therefore, the intelligence was used to keep these animal creatures, and the other things that would not allow these groups to progress and to grow — they were more or less controlled. "This One" was a controller and through mental images and mental powers was helping to control this situation, so that the young ones on the Earth were able to develop without quite so many obstacles.

As this time continued, "This One" felt that it would be time enough and that they should continue on their own. She had completed her work and there were other things to do. It was well understood that there would need to be ones that still were of a vibratory frequency put on the Earth that would be able to be connected — that had not lost the correct vibratory frequencies as those who were transformed into that physical body during that period. That it would be wise to have ones that would still be able to contact me. "This One" was chosen for this project and also so that she may learn the understanding of people. People must be guided and still have a connection to where guidance may be given. "This One", as many others were also, was chosen to do this and that is why "This One" is on Earth. Is that understood?

Not totally, but I think when we listen to the tape again we will understand it better.

Continuance on with this searching and this looking has become quite a quest for this knowledge. This is why I told you before that the looking for the God-head was more a wanting to get back to what was, not wanting to get back to a God.

Then will there be a time when we totally leave this Earth? Is there a way that we can get back to what we once were?

You will never again be what you once were. As all progression continues.

Could I help my conditioning by meditating at a certain time?

At this point do not be concerned with that. This will be regulated and this will be taken care of from this part. You continue on with your meditation.

That is all of the questions I have for now. This knowledge, I'm sure, will open many minds.

Chapter Three

(Tape 13)

You said that this Earth is slowing down. Is there anything that you can tell me about that?

About the Earth slowing down?

Yes, I guess what I am asking is that our scientists have speculated about the Earth being developed from a "big bang" theory. If that is true how does that fit in with the slowing down of the Earth's vibrations?

Into this "big bang" — exactly the way it is understood or explained is not correct. Your Earth was formed and put together and created as stars or as other pieces are. This was not exploded. The frequency with which the Earth spins, or is spinning, or has spun before, is all within the making, within the correct proportions. It was as molecules, and as pieces and gasses were all forming and coming together. Winds, and the creation vibrations were making it go much faster than it is at this present moment, as was necessary to be formed.

As it is slowing down, and as it has formed and other elements are now beginning to take hold and to reshape itself, vibrations are changing and new elements are coming into creation. This all changes vibratory frequencies. All change, as has been described before. Everything changes and vibratory frequencies continue to slow. This was the process. This is the process for the Universe to take its natural course. No one was controlling this.

Then this was just a natural thing that happened to the Universe?

This Universe was created in sense and form as explained before, but still within natural laws, and not created as a person takes a piece of clay and makes a sculpture from it, exactly. As your understandings are speeded up, and as basic understanding is given, then understanding in more depth may be given. It is very difficult to explain to you in terms of universal working and all Universal Laws. The explanation that is given would have no meaning for you at this time. As time goes on I will get into more depth with these things.

Linda, under hypnosis, can see the aura much differently. In reading books that are here we can't find anything about what she is seeing.

Understand that when the conscious and the sub-conscious are two separate entities, the sub-conscious is what you were and the conscious is what you are now.

Then you think in her sub-conscious is the best way for her to see these auras?

She does not see them fully as of yet. Understand that she will be given this slowly, as she can understand. Do not push too hard as this is a natural thing she is trying to develop. This is not an unnatural thing. Also this has not been practiced by her for many lifetimes and even in her sub-conscious this is being retrained and brought forward and relearned again. The better she understands it, then more will come forward, as in your training as I have explained to you.

Then at this point you are telling me that the way she should see them would be in the conscious and the sub-conscious?

She must work with both if she wishes to see both. Much help will be given to her and put into her conscious mind also. That way both may be brought about and much will be gained in both while this is being done.

We would like to ask some questions to better understand ourselves and those around us, including our children. If we are not going back to the source, where are we going? What are the lessons that we are learning for?

I have not said that you will not return to the source. You wished to know if you would return to what you were before you came into your Earth's existence. I said "You can never return to what you were."

Then are we going to other planes of existence, eventually when we have learned our lessons here? We wonder what these lessons are for. When we graduate from college we feel we will go to better jobs, how would that apply to what we are doing on Earth at this time?

In a wider scope, and a much more undefinable way, it is basically the same, as the same thought continues in a different pattern, however.

Can you tell me about the different pattern?

These questions that you ask are asked of one who is not of your plane, nor of the dimensional effects of your Earth or solar system. When questions are asked by people, they are normally answered by those who are on your plane, or who have been there, or that are within your solar system dimension. These questions are then answered with the understanding that is from that dimension or that time period. However, this is not the case and things must be defined more clearly. How do you wish these answers to come? Are they to be answered strictly from the Earthly understanding? Or from the understanding of those who were of the Earthly plane at one time and now are in the Earthly dimension, such as your guides? Or do you wish answers that are from other dimensions and universes outside your Earth, from where I am? This becomes quite complicated for your understanding, and does make for some confusion. For when you ask, it is hard for me to explain all. You have asked for universal knowledge, therefore, how do you wish these questions to be answered?

Well, I think probably they should be answered from the knowledge of the ones who have been here and gone on before us (such as guides). If you answered from the universal knowledge, I'm not sure I would understand. Yes, I think that will be fine until we become more aware, then we can get deeper into the answers.

That will be acceptable.

Thank you for giving me the choice. Now I think I will re-phrase my question. What are the lessons we are learning on Earth going to have to do with our progression? Like people on this Earth praying to God to return to him.

To return to God is a very simple matter, but this will not be the end. To return to God or the God-head will only be a small portion of this, so striving to do this is good from your plane. I just wish to make this understood.

But I think people want to know where our lessons are taking us.

At this time, with the understanding on the dimension you are now on, this will eventually return you to the portion that is thought of as God.

Can you tell me more about this portion that is thought of as God?

This portion that is thought of as God, and the beginning, and the creator, is but a small portion of the whole. However, it is known to you on your plane and dimensional effect. This is still, as I have said before, from the source and from the beginning, as understood from the Earth's dimension. The wisdom to want to get back to that source is of good wisdom, and good thinking and planning from the Earthly viewpoint and from that dimension, while you are in that dimension. This is the main process that you will be trying to achieve.

Is that all?

No, that is not all but it should be sufficient at this time. This is of the Earth's understanding at this time period.

I guess what I'm trying to do is to go beyond the Earth's under-standing. I feel that I know the part that you just told me. What I'm looking for is beyond that. So will you give us the understanding from your dimension? I think that would be the best way.

This is as trying to achieve learning of one important step or of one phase with one particular teacher, which you call God. All aspects of this teacher are within the immediate solar system, galaxy type of division. These lessons, learned well, from this one source, from this one teacher, will be of much importance and much learning and much advancement, however, this is not the end. To go, to link up with, to have the same understandings as this one teacher is not the end. As it would be in your school, you would go to another class, to another teacher, to another grade, to another University as you progressed.

As I have said before, much in the dimension and the Universes is the same type of a general pattern that is within all order. Although, methods and things desired are different and are brought about in different ways, and are presented for different learnings, it is much as I say, as for one teacher. To be able to be in tune with this creative source, as the creative source totally, is the goal for this one experience although it is from your portion of understanding and learning many, many varied and dimensional learnings. Therefore, this is of importance, and generally the way it is conceived, is of truth and is the way it will be, although, it is only a small portion and there is much more.

Is there anything else you could give us along these lines, that you feel we could understand now?

Try to understand what has been said. The universe and beyond the universe, the dimensions are vast and varied and very large. What you seek now, from your level, is almost unbelievably difficult and much is to be learned, and it is very hard to see how you will ever learn all this. And yet, when viewed from where I am this is but a very small portion of your learning. Learn as quickly and as well as you can. Learn from where you are, then you will go on to other learnings.

That's what we are doing now. It's just that not all people believe that we are returning to a God source and then that is all there is to it.

53

That is true from the basic understanding of this Earth period.

O.K., but the ones that choose to believe that there is something beyond that?

Go back to what I said in the beginning. Understandings have been given from other planes of existence that are from within the dimension that you are on now. Therefore, the source is the most important, the top of the chain, the end. To go back to that source is the understanding at that dimensional level. Those who believe this and understand this have an understanding of that one existence of that plane and dimension.

However, there are those like you who question this. There are not any, as has been stated before, who are capable of coming through into a totally different dimensional effect except "This One". At one time there were. Now, through not using this ability or being lost on the more Earthly dimensional planes, and not using the ability they have had, this has been brought down by the Earthly vibrations as it was in the beginning and therefore has been lost.

The dimension I am working from, and am in, is not often reached. Therefore, the understandings and the knowledges that are being given are given by guides from the Earthly dimensional planes. You ask for higher knowledge and will receive them. So you must understand from where knowledge has come from and now where you ask knowledge from. These are not the same.

It is good for us to understand more about where you are in your dimension and your universe.

Understanding of my Universe and dimension is not possible for you now until other things are learned and closer dimensions are understood by you. Then you may understand more of where I am and what happens here.

That's fine with us. We have some more questions. We were wondering about love and hate and dimensions outside of this dimen-

sion. Is there love and hate, as we know it on Earth, in the Universe too?

Are you speaking of when your body discontinues?

Yes.

Emotion is both of psychic and animalistic in nature, and combined. When combined, they become a different entity, although still part of both. The compassion, understanding and feelings come from both. The severity of rage and of jealousy are of the opposite end — far away from love. These are more of the animal type reactions. To get the emotions back to a more stable, normal understanding is part of the meditation and the conditioning that is going on, on your plane, to bring these things to a more stable understanding. So there is not such emotion as hate, or that there is such emotion as love only through sex. But so that there is compassion and true love and true understanding, then there are not such extremeties. This is part of the training and learning that must take place.

Are there men and women on other dimensions too, or is this just an Earthly thing?

There are, on this other dimension that you speak of and ask questions of now (after death), still feelings of male and female. There is still brought forward with you the emotional tie that will make you male or female, although, this was not necessarily true in the beginning. These energies were combined and worked through one soul entity, but as taking bodies into male and female for the Earth experience, this began to separate one's thoughts and growth patterns into the two different divisions and, therefore, this feeling is continued much on the other plane.

Then our souls, at this time, are also being changed in some ways from what they were through our learnings, our feelings and vibrations?

That is correct.

Then this will continue from now on?

As I have said, there is no going back to what you once were, as all is progress within the Universe. However, certain changes, and certain things that are changed for this experience, may only be for this experience. After this experience is totally over with, they will go back the way they were. In some senses, and in some beings, perhaps not totally, but they will go back to the original form. As you grow from a child to an adult you are still the same but actually different because you are no longer a child. Is that understood?

While you deviate from what you were when you came to this existence, and you make changes to adjust to your experiences, still you will go back to the basic original. Like male and female, when this experience is totally over and male and female will be more balanced, and more harmonious and will be more like it was. There will not be the definite separation as there is now.

Then this is just one phase of our learning, this entire experience, from the source, or God, to the returning to the source, or God? This is just one continuous learning session?

That is correct, and not all learning.

Then if we learn our lessons faster through discipline, or whatever it takes, then do we return quicker to the source?

Yes, of course, you return quicker.

Then that allows us to go on to other learnings in other areas? Whatever they might be?

That is correct.

Are there areas that are beyond this area of learning that we are now on? And what about the area you are now in?

Where I come from, and where I work from, is not only beyond your learning area, it is also in a totally different dimensional plane and Universe.

Will we ever go there as our learning progresses?

Certainly you may.

Good, I would like to meet you someday. I think "This One" has already met you.

"This One" is originally from where I am now.

Is there anything you would like to tell her about this?

When the time comes I will discuss this further, as stated.

O.K. "This One" wanted to ask you about a dream she had about a person being a certain color and discussing about Vitamin B-Complex. She wanted to know if this had to do with her aura training.

Of course this is part, and through her dreams she will be helped. To remember these will be of importance. But this was also of twofold. This was also to remind her that this is needed, as she had been wondering if she needed more B-Complex. This was to tell her, "Yes, do be sure to continue on with this and do not let it slip by." Also it was, while giving that message, to allow her to know that this will be, when she sees the aura in this certain way, that this will pertain to this one thing.

I was curious about something that you said the other day to me, that much help had been given to me.

Much help has been given in many ways that are not realized by you. However, there is more also. As you know there have been many times through your existence that you have needed much help and it has been given to you from other planes and other Earthly existences also. Your life has been one that has not been easily contained. This is an understanding you have already made. The understandings that these feelings within yourself and the protections that have been given to you are for a definite purpose that you are to have in this lifetime. It has not been easy.

I truly understand, and as I am understanding, I am making adjustments.

This is very good. Much progress, within the understanding and within the realization, and then acting upon that, is then made.

On this plane can you tell me anything about our memory. Is the memory actually stored within our body or is all memory stored somewhere else that we draw from?

Memory is a very intricate part of your being. It is of emotional, it is of mental, it is of all bodies that are within and contained and used by you. Each portion has its own definite function. Each cell of your body is a part of the network and has its own particular memory and understanding. But, the memory that you wish to know of, and are concerned with at this point, is the memory of the brain. Of the portion that is utilized to remember Earthly things and to remember what is happening now or to remember phrases and words. This physical memory has a large part to do with the brain, with the nourishment of the cells of that area, and can be trained and can be retrained and controlled upon mental exercises if needed. There has been much discussed, and much given, and much understood about memory and memory training.

We will close for now. Go well.

Chapter Four
(Tape 14)

NOTE TO THE READER: As you are reading each one of the chapters in this
book, please keep in mind that the
knowledge in this book (and more) has
been coming over a period of 2 years.
You will find the questions may be
asked more than once, or may be
asked in a different way, to bring out
the full knowledge for your understanding.

Do all souls come to this dimension, this particular cycle?

No.

*Is there a reason why certain souls go a certain way and other
souls go another way?*

By choice and by knowledge at the point they make the
choice. It is as certain grades have certain classes they may
choose from. The conscious learning ability only has to do
with the physical component on the physical level at this time
period. This is of the physical time period only. This does not
mean because one in this time period, this Earthly experience,
does not have much education, that this soul does not have
any wisdom. Wisdom with the soul, or understanding of the
soul, is not manifested necessarily, in full understanding, in any
Earthly time period. Divisions and understandings of learnings that are given in each instance has to do with that lesson
being learned. It does not necessarily have to do with the col-

lective understanding of the soul. Judgement, as has been stated many times in all works of great literature on your Earth plane, is correct when speaking of judgement of others. While it is evident that the judgement may be made from a physical level of seeing and understanding, that has nothing to do with the soul's development and understanding.

This is as watching someone in a play — as the character is at that particular time, the role in the play, so to speak. Then you judge what that character is, how he acts, what he is, and how he looks, and this is your opinion of that particular character in that play. However, when not in the play, that person may be totally different, have nothing to do with that character.

This is as we are seeing things on the Earth plane. Try to understand it in that way and you will be able to understand the dimensional effect better. As in your motion picture, one person may play dozens and dozens of different characters but this does not have anything to do with the real actor, the real person, his real life or his real intents. Do you understand?

I think so. I will go one step further. Things that we consider sins here, do they have anything to do with our real being or are they just part of our role here?

Have understanding that sins differ in universal mistakes and man-made understanding mistakes — mistakes, or sins as man has deemed them. These are truly of negative kind whether man-made or Universal. Even if man-made, if understood by the person as a sin, or as something of that type, it will be of detriment to that soul or to that understanding person. Therefore, it is not recommended to sin. While it may not necessarily have to do with God, and punishment and so forth, as understood there, it does have to do with going against one's own natural development and unfolding.

Many times these negative things are brought into being for learning and understanding, and from these do you grow and do you understand, always. Always these are for growth if they are accepted for growth and lessons learned. As said before, the situation can be repeated over and over again until corrected. Not that the situation changes **but that you change,** and then can handle the situation. That is the purpose of negativity or sin, however.

I always thought the biggest sin on Earth was to hurt another person.

This is very negative. To injure another soul is, of course, the most injurious thing that you could do. To hurt another, much, much learning must come from that. There must be much learning when you hurt another or the karma that comes from that, again is very negative and much learning will have to be given to be made up for that.

I've always felt that I was being kept alive for something, but I didn't know what. Am I scheduled to do something besides my learning?

That is the most important. This is being fulfilled. Much learning is to be given to others, and as I said, this has not been easy because of the fact that there is not that much learning and understanding that is coming beyond the Earthly plane and Earthly dimensions. Time is, as is known now, that other understandings and other truths need to be given and therefore, this is being done.

Thank you, I will do my best. When we opened this door into the Universe did I do it myself or was I guided into opening it? I was always curious about that, if my curious mind took me to this or —

Without your curiosity and your curious mind, how could one guide you?

Then I did have help?

Help has been given to you since the beginning. Guidance has been given to you to fulfill this need that is now. But without your curious mind and without your striving it could not possibly have been done. Realize also that there are reasons for your coming into experiences and guidances. Very simply put, and although I understand it is a very difficult science, astrology does play its part. Astrology is for a purpose and for a reason and therefore examine beyond your own self, why, when and what. Your purpose, your experience, your guidance, your desires have all been brought together through series of understandings and knowledges. Even before your existence onto your Earth time and plane, it has been prepared. As known and understood that while guidance is given, still free will may be used. But still, the biggest part of lives is given as is arranged previously although many changes may be made, and normally are made. But still, there is a reason and a pattern for your being and coming into being for your existence. If this is not met, through free will, and is not completed on this time period, it will be given again, and similar circumstances brought about to make this pattern, until it is completed.

If we knew on this plane what we were here for, it would help us to complete it, right?

It should speed it up, yes.

Is there a way of knowing some of the lessons that we come to learn?

Understanding past lives will help you to see and to understand and have a pattern of what is happening to you.

Does having possessions have anything to do with this?

You may have things and may acquire things and may enjoy them but you must not have possession of them. They must not control your life and they must not be the most important thing to you that there is in your lifetime. While the lesson may be learned in one lifetime this is not necessarily true. It may take many to break totally away from the desire.

Is it good for us to get a guide on this plane?

If you wish a guide to communicate with you from the plane and the dimension and understanding of that guide, then you may contact your own guide.

Will those guides only give us information about ourselves?

Those guides have information that they can give you on the level that they work from. Many guides work with people that are only on a level not too far above the understanding and comprehension of the person they work with. While other guides come from other planes of existence or dimensions to help. Therefore, you only receive the knowledge that the guide can give you at that time. Dimensional learning is the understanding that is passed forward — as a teacher that only understands 8th grade level teachings can only give the teachings through the 8th grade. Is that understood?

Yes, I understand that a guide's knowledge can be very limited.

Different guides can go to different levels.

Will they put us in touch with different guides for other learnings?

They can, yes.

There are questions we would not bother you with — perhaps if we could ask on these other levels.

You have not been given help from these other levels, as a guide would give you.

Yes, I understand that we have gone far beyond the level of a guide.

If you wish to contact "This One's" guide you may contact Marcheah as already known. She is a very high spiritual guide and can give you much help from that dimension if you feel that is necessary. However, she continues to work with you known or unknown, it does not matter. As has been said before, understanding will be coming more quickly to you all the time. Acceleration has begun and will continue as knowledge is given on one plane or on one existence also, then the foundation is laid for knowledge from another dimension and greater understanding will be given as you can see from the time we have spent together.

Yes, I can see a big change in my thinking in a lot of ways. Will my conditioning increase my psychic abilities and understanding of people?

Much understanding is had by you of people, much understanding. Do understand that the knowledge and training you are now receiving is far beyond what you asked for. Accept and understand it. All that is learned is put into the being that you are becoming. All that you desire is of importance to continue on different levels. Therefore, learn all you can of what is of interest to you. When the time is right, all comes together to make what you become.

People often wonder why this information is coming through us, since "This One" and I are not famous.

This has nothing to do with the dimensional connections. It is not necessary for "This One" or you to be of great importance on your Earth plane. That is of no importance to us. Understandings and teachings that are given to you directly is of great importance, and your station on your Earth has nothing to do with this. This is totally separate.

We were surprised at the time element when this did come through. This was to have been sooner, but was not and, therefore, we are happy to be able to make this contact at this time. Do not underestimate the value and importance of the conditioning, and it should be continued.

I'm talking about when I talk to other people and I explain to them where these thoughts and ideas are coming from.

Many can contact their guides, many can contact those who are of very high dimensional effect in your universal dimension there. However, at this particular time there is only "This One" who is in contact with this dimensional effect and with this Universe and can bring in the teachings for Earth, and this, therefore, means more attention is given to her. But there could be many others as this was set up at one time.

Let me put it this way, why are we the only ones who can get these particular teachings?

Because "This One", through her incarnations, has not allowed this to be brought into a mundane level. "This One" has only continued, through receiving and sending to this dimension, and has not gotten into the habit, through previous incarnations, of using the knowledge, and what she could, for things that are on an Earthly level. There have been many, who knowing and understanding the abilities they have had, have used them and have only seen and been able to understand what was on the level that they worked on from that time. Therefore, the vibratory rate was eventually lowered and many of these people are still in your Earth plane and are even incarnated now, but they have lowered their vibrations and the way back to my Universe will never be regained.

Many on your level are clairvoyant, many can seek and understand things that are different than the normal on your Earth's plane. "This One" has not used this knowledge through any lifetime for those reasons. Although, the feelings of the power to do so have been there. This way the channel has remained open through "This One" only.

Can you give us more on the Earth changes that are coming?

There are several different various levels of teachings and understandings on the Earth's plane. Outside of the Earth's dimensional effect there are other teachings and other knowledges. Outside of the Earth's dimension is where I come from and this is where I give teachings from. This time period is becoming very critical for people of the Earth. In your great works of knowledge, it is known that each time there is a need, that there is someone that comes to give aid. These teachings are now being given from my dimension for your understanding.

Be aware that your Earth is a being. It is not a thing. It is changing and becoming something else as previously planned and is within its own development. As this development changes and continues forward, all must go forward with it or all must go elsewhere. Vibratory effects are of your whole Universe, and beyond your Universe, and vibratory effects of your Earth are being changed. These must be changed by the soul developments also. The vibratory pattern of your learning area is being changed and those who cannot match these vibratory effects will not be able to enter into your existence more.

These soul developments will not be lost or doomed in any way. They will simply continue on their learning experience somewhere else where their vibrations do match. It will be on another plane, on another place or another planet — it will just not be on Earth. Therefore, the vibrations of your planet, of the people and of every living thing is being changed and raised. There is not a way that the vibrations of plants, animals and everything that is living to be raised and not your vibrations be raised also. They must be raised to match the others.

This is not cutting off of those whose vibrations do not match, they will simply be sent somewhere their vibrations will match. Much help is being given from many different areas without direct interference. As we talked about your children, they may be guided as we are guiding the Earth and its children. But the Earth, as with your children, may not be interfered with directly as stated. Guidance and help of all kinds may be given, this is all.

As time goes along, will I be able to get any information from your dimension as to catastrophies that are going to happen here? I know some people get information about earthquakes. Will I be able to get that information?

This is not my intent.

I just wondered if it would be possible for me to do, maybe even get it from guides from my dimension?

This is possible for you to do. But it is not my intent. One thing that I feel I would like to say. Try to understand this. While being as Guardian for your planet, and while trying to guide or to give information, along with others, as I am not the only one, I wish this understood. When we observe, or when we watch, or absorb knowledge from you, or give knowledge to you, the things that you have made your own things of importance are of no importance here. The money, the status of your Earthly life, the color of your skin, the working of your government, other than the effects upon all, are all of no importance. The things we see of importance are of the vibrations of the individuals and of the soul developments that are coming through. Importance of these things is great to us. We are not concerned with others. These are your thoughts and importances, not ours. Understandings that a soul has, and a soul's development, has nothing to do with who they are, or what they are, or where they work or what kind of car they drive. These things are of no importance period! They are only of importance to you.

Several questions have been made by you. Would you care to ask now?

I'm sorry I can't honestly think of what they are.

One question you have had within your aura that seems to need some clarification, that I would like to give, is a question of the two souls.

Oh yes, thank you.

Of the souls that are within the body, within the understanding of when the soul was taken into the body. Is that correct?

Yes, I've been wondering about the conscious and the sub-conscious and the combining of the two or the separation of the two.

Very simply I will give you this at this time period. When the soul, the intelligence, came and was watching, there was the animal soul that was within the animal body. There was also the intelligence of that animal intelligence. When the adjustment period came and when the agreement was reached, then the finer bodies that combine and that make up the soul, as it is called, spirit, as all of this was combined and then was combined with the animalistic body to house it or to house a portion of it, as it does not encompass all, understand this; it is only as a base of operation type of thing, there is only a portion that is enclosed within the animal body, this intelligence of the animal was then raised, was changed. Then, when inhabitation occurred, then the two were combined. Therefore, you have the thinking, feeling, conscious mind that is part of the animalistic body that is there, that is working to take care of it, to help it in everyday life. Then you have the sub-conscious, that is called, that is taking care of and working along with the conscious mind and helping to regulate the other bodies that make the soul development body. That is the electrical part of the body, the mental body, the emotional body. All of these that were incorporated within themselves were incorporated with the animals' bodies of that type also. Therefore, you have a blending of the two while still you have the conscious and sub-conscious mind that is trying to work together to control both and also the parts that have become one.

Therefore, you have the conflict. The conscious mind of the animal is necessary to be in control for the conscious part of what you do for every day. The sub-conscious mind is in control of the part that you are not aware of for everyday use and also not even aware of at all. For as many teachings as have been given, there are still more that are not known of the bodies and the bodies that have been created by the two being combined. These knowledges will be given at a later time and will be given in small portions so that it may be understood better.

The sub-conscious, which is in control constantly of these other finer bodies and the conscious, which is in control of the grosser bodies that you see and work with everyday, then these two must combine. Much of this is done during the sleep process where your physical bodies and your other finer bodies are either at rest or they are doing repair work and they are merging and this part that is controlling the two that are merging, and have merged, is more of the dreaming part.

Dreaming, as told to you earlier, is much more than is imagined. Much more than is even thought of. Control for your daily life? Yes. Help for understanding on the health level? Yes. On the understanding of "Are you learning your lessons?" Yes. Also, this is being worked together to help the bodies that have been fused and are trying to be controlled. More on this will be given later as can be understood.

However, these two minds do combine and do work together. It is just that the conscious mind has been given free rein, because the sub-conscious mind has been lost through many centuries. It has been forgotten and, therefore, has not been consciously brought about and the two minds been put together. Yogis have some concept and understanding of how to bring the two into an adjustment and into a better working area so that the bodies may all be put into a more harmonious way, but there is much more and this will be a large part of the learning and understandings that will be your lessons there in your dimension that you work from now.

Many teachers come to teach you individual lessons. Individual thoughts of love and unity. Jesus, Buddha, Mohammad and others taught many individual things. All leaders teach these things and when basically viewed are the same, just in different ways. As the smaller lessons are learned and you learn to accept their truths, and when you learn to put them into **practice,** and use them to control yourselves, to bring your bodies together and to merge them together again, so that control is total on both sides, conscious and sub-conscious, then is when you will be able to use them together or to separate them apart and continue on.

So you see, it is not only one thing. It is not just one avenue that must be sought. There are many small things that must be learned and practiced that then go into the larger thing that must be done. All learnings, all teachings, are of very great importance, and control is very important. Vibrations are the essential thing. Do you have any question on this? Perhaps I could make it more clear for you.

When we do the book on the soul, we would like to discuss it further. Is there going to be a time when we will not use this Earth as a place of learning?

At a time when your Earth will not be used as a place of learning?

In the beginning, when we came here, it was not considered a place of learning for us.

In the beginning, this had its developments and its purpose, and at that particular time the consciousness which came into being was not prepared for that, no. This was as stated before through man's — no, not man at that time — but as the soul

69

that came into this place, it was through their free will and their desire that this was brought about on your Earth. This was permitted to continue as was desired by them at that time and was accepted and was looked upon, more or less in your terms, as an experiment.

This has been watched and guided and guarded carefully as in most realms of creation. Everything in all the dimensional effects are more purposely designed. Therefore, this plane that you and many have chosen has become a learning place that is not exactly what would be considered a normal learning place of the Universe. Much learning is taking place there. That is why you are told to learn and to do all you can there.

Your Earth has its own unique factors. Free will, that has been given everywhere, is very abundant there and much learning is to be made, and much advancement for the Universal knowledge is being given and being put into use there. Knowledge is accepted very slowly and understanding comes very slowly, not only because of free will but because of vibrations and other things that are difficult to explain to you now.

The Earth is a good place of learning. Do not be ready to rush off until your learning has been completed. You can leave without total knowledge, without full learning, which is of very sad consequence.

I was asking from the more Earthly plane, because I know La Mureah and Atlantis were both destroyed and maybe there is a lot of learning from that.

That is true. Much learning is done through this. But this is all part of your acceptance. As when children are born to parents, they accept part of their karma, and as you marry and have children, then agreement is reached that you will share, and you will help, and you will accept part of their karma.

It is also the same with sharing of the Earth, and the sharing of this particular dimensional effect, it is agreed upon. These are all workings of togetherness. These are all learning experiences of sharing what is happening to the other, and instead of looking upon these things as disasters or as hardships in your life, they must be looked upon with more wisdom. They must be looked upon as experiences of sharing and great wisdom and joy. Even though they bring you more heartache and trouble, these are still all learning experiences and to learn this and to accept it in your heart and soul, and being able to share and to respect each other's dignity, and each other's problems and accept them with graciousness, love and an open heart, is all a very large part of your learning. This is a positive learning.

The negativeness of complaining and looking at the bad side of everything continually, makes even more problems. Learning to love and accept and share openly and happily without sadness — this is a very large learning. If you learn this, you have learned a great deal. Each thing is an individual thing, but is also part of the whole. As a drop of water is a part of a glass of water, and a glass of water is part of an ocean.

I think each time the Earth has gone through big changes like La Mureah and Atlantis that new and varied understandings arise. I believe that since we are due for more Earthly changes that this is why new teachings are being advanced.

Your thinkings are correct.

In the beginning we made some agreements when we came upon this Earth, when we agreed to take only the body that houses man at this time. I would like to know what those agreements were.

Agreements were made with the other entities that were there, with the soul developments of that time.

The soul developments that were here at that time? Are your referring to animal souls?

All soul developments that were present on the Earth at that time. That does include animals also.

Can you tell me a little more about the souls that were there at that time?

Varied souls, varied kingdoms upon your Earth. All living things have souls, as known. All have their own development and all are on their own path. Intrusions that were made through the desires, as previously stated, are still coming through and being worked through.

Did any agreements that were made at that time have anything to do with our free will or you not being able to intervene directly?

This does have to do some with this, however, it is not all. It is not the place of those who are watching or guiding to interfere with the development and the progress that is made from any level. There are times when interference is made. There are times when things are done that are not of a higher nature. Entities that are within the realms of the Earth often are anxious to help and are anxious to cooperate and to prove their existence or to prove the necessity for their help. Interference is then given, and then must be met on the level from which they are working at that time.

Much the same as interference, as with one another on the Earth, at the time that help is asked to be given. As always, things are basically the same throughout the different worlds and planes, however, the entities of vast knowledge, entities on a level that understand from a higher and more technical degree, understand that interference is not to be given.

There are many people on our Earth who have reached guides and other entities. How would this connection differ from theirs?

There are many that are working on the level that you speak of. There are many who are doing what is necessary to be done there. However, the connection from you to here is of great importance for the type of thing that is now being done. One way connection can only be done through the talking or the exchanging of the thoughts for the guidance. Is that understood? The connection that comes from you to this place here, then makes the connection strong.

Also, do understand that this is not only of mind. This is also of spirit and of direct vibration that is not only being met where you are, but is being met at another place in another way. This is not only as one coming through to someone there. This is being done by a two-way communication that is being communicated at another point of distance. Those that are at your level are still not capable of this type of communication. As a satellite circling your Earth and as the beams of the television go to the satellite and back to another distance on the Earth, there is this type of communication being given here. This is the difference. This is the main reason this is not being done there.

Also, with the type of things that are being done on your Earth plane, such as guides and other entities, I do not spend all my time working on this level. There are many who are doing this. There is not necessity for all to do this all the time. Full devotion is not given to your Earth only, but the type of communication that is being given here, at this time, under the condition as stated before on the voice tracings that you are taking, is being given and is being done. There is no great mystery here — perhaps that is not correct — there is no great mystery from my level, you see.

This is very true, but from my level it is a little different.

I understand. "This One" comes to me at a point beyond your galaxy, into your Universe, beyond where you can understand at this time. I come from where I am and go to a place where we both meet. My time, my space, my understanding is altered drastically, as is hers, but mine is altered drastically as I am communicating with you, with her and being replied and relayed back to you. Therefore, I am completely away from my time and space as would be thought. The meeting here is as a place we can both handle well. Does that answer your question or do you wish even more?

At this time that is sufficient. Should these new concepts or teachings be added to our present method of soul progression, or should they be on a separate path, given separately?

These teachings that have been started, and shall be given in a more grand scale as time goes on, should be added to what has already been given that is known. However, many will differ slightly and many will be changed. There will, of course, be some conflict "Is this teaching correct?" "Should I adjust my thoughts to the new direction into which these are going?" Do change your thoughts. Do listen to what is said.

Do not put into it what you think it means. Listen and do not project your Earthly thoughts into what I am saying. Listen to the words and then listen to what is stated. It will be given clearly. If it is not given clearly, do ask for a better understanding for what is said. Do not put this aside until it is understood.

Things will be changed. Things have been given in the past on a level that could be understood for that time, or the teachings that were given were changed and misdirected by those as they understood them at that time. These teachings will now, many of them, be corrected. There will also be new thoughts and new ideas interjected, as stated, they will not be of earth-shattering proportion. I am not bringing a new religion into your dimension. I am not bringing any religion into your dimension. Nor do the teachings I give have anything to do with religion. Do understand that. Religion is a thing that is made for man's desires and purposes.

These will be teachings of laws within the Earth plane. They will be teachings of laws that go beyond the Earth's plane into a more dimensional effect of Earth. Then there will be teachings that will encompass your galaxy, solar system, universe. There will also be teachings, towards the end, once the first teachings are understood, that will begin to give, to the populations of your Earth, the teachings and understandings of other universes and galaxies. These teachings and understandings will be on a very small plane from my viewpoint. This will not be high teachings of my Universe, although to you they will be very high teachings. These must be started to be interjected into your society and your people's thoughts so that they may begin to understand on that dimensional level. When they begin to see those teachings and understand them, then it will be time to give to you higher teachings and understandings of my Universe and dimension.

These teachings will constantly be given to you more and better for your understanding. It is time for understanding to be brought forward. Your universe is vast and the teachings that have been given to you have been kept down — "tight" would be a word to use I think — but it has been kept in very small ways and has not been expanded correctly and your technology is growing rapidly now, and, therefore, your mental processes are growing more rapidly, so it is time for your better understanding. You cannot continue advancing yourselves mentally with the understandings that have been given to you many, many years ago. This is not correct. This cannot be. You must grow and advance spiritually and understand yourselves as a spiritual being. That word, spiritual, is not correct. It is not a good word, but it is used at this point so that I may communicate an understanding with you. This will be discussed later. There are other terminologies that you must become aware of that will be more appropriate for your time.

Yes, I realized that you were not of a religious nature. That we will be learning fact, of how to change our vibrations.

That is basically correct.

I was trying to get it to the people this way, but it is very hard because there has been so much misunderstanding about Jesus' teachings. Well, he is the one most talked about.

This is the one that has been continually spoken of in your time. This must be expanded. This must not continue — to allow your mind to dwell upon one teacher. This must be expanded to other teachers.

Are we the only one that is responsible for our soul?

This is the only way it can be done. Another cannot help you. Another cannot do this for you. This is through changing of your own vibratory patterns. This is your own desire. This is your own effort that is put forth. If others could do this for you, there would be much help in the world, and in the Universe.

You are an individual unto yourself, within your aura, within your own field, within your bodies even, this is your control. You are the controller of much. Within your own self alone there are millions and millions of entities — no, this will not convey the correct meaning to you — bodies and developments that are within your body and your control. Within your aura you are the master of all you do, and have, and are.

There is only help given by others of suggestion and physical help if you are in need, but all must take place within your

75

own body, within your own aura of mental and emotional development.

In another discussion we talked about our awareness progressing us faster. I would like to discuss this further.

It has been stated before that with awareness, development does come much quicker. You may do something in your lifetime, with no awareness of progression into what you are doing, and development and progression will be made. However, when the time comes of awareness and understanding, the same thing that is done will produce much more development within the soul. So, awareness is of great importance. Those who know what they are doing and do things for a definite purpose will be advancing themselves much faster and will have much more benefit than someone who just stumbles into doing something correctly.

How do you convey to enough people, to help more people get on the level of awareness that they are seeking?

One way is of talking and projecting yourself to others, not only talking but of the vibratory field that is portrayed and is intermingled with others' vibratory fields. These vibratory fields will convey much and will progress much, even more than spoken words. Words contact the conscious mind and put a thought and a feeling into a person's consciousness, and also into the aura goes your feelings. Communication and inner-communication is done that way. Doing what you believe is a very good way of projecting yourself, also of speaking and of doing. Much help is given to those by talking and — gentle persuasion would be a good word, I think.

Some people on this plane think that if you go to church and pray to God, that that's it. It is my understanding that if they do not stop eating meat and do not start meditating and do a lot of other things that we have talked about, that praying to God will not change their vibrations and will not take them to where they want to go.

What you say is true — this is basically true — however, there is much to be said for what you are saying also. As I have said, praying to God is important. It is a way that many people can control the emotion and have understanding of themselves. This is of a personal thing. The feeling of communication with a God is of much value to many.

The eating of meat is — you ask of the agreements of the beginning. The eating of animals was not agreed upon. This was not originally within the thought even, that this would happen between the developments of the souls of the animals that they were coming in to be with and live with and to communicate with. The eating of the animal kingdom is not good. This has been stated before, but many people are not of a vibratory frequency that they may understand this. Those who are of a higher level, or who are on a level who wish to refine their vibrations, who wish to have the communications and understanding, and more peace that is desired, understand that this is not a thing to do.

There are many levels from which people can desire not to eat flesh. There are many reasonings that they give themselves on the conscious level, but it is on the sub-conscious level that this desire is there and this understanding is there. This is being brought forward more. The more people that have the understanding the better.

To simply stop eating meat will help the vibrations much, but as stated, knowing why and understanding why will help this to refine and attune oneself much faster and will be of great help. This must be of their own accord and their own desire, their own inner understanding.

When we are talking about animals, they have souls too. When an animal is killed without making his proper progression, under his normal patterns, does this interfere with his progression too?

Look to yourself, there is no difference.

In other words, because a cow has been raised for the purpose of slaughtering, they still have their soul progression which is being stopped, right?

That is correct.

Then it's more than the eating of the meat. It's interfering with the way things should be on this Earth?

That is another reason, yes.

I will close for now. We have enjoyed the time spent with you this day.

Chapter Five

(Tape 15)

In discussing meditation with my family I've been trying to convince them, and explain to them, about meditation.

No need to try to convince. Do not try to force. This should not be done on any level. Those who have been alive in your Earth's existence and are now deceased, and are able to communicate with those who are alive now, these souls have the same understanding that many that are alive now have, except some are advanced in some areas. In the other dimension, things are somewhat different and changed there. Do not expect all things to be of a high level that you would expect great knowledge from. Many are willing to help and talk and guide from where they can. All feel they are helping. All knowledge that comes from that dimension is not of the quality which you seek now or which is needed to be given into your dimension. Therefore, influences from there that are made can be interference. Do not mistake something for guidance that will be interference.

You must look closely at what is given. It is also the same on your plane. Look closely at what you give. Are you giving guidance or are you giving interference? If you are giving guidance, you will guide and then you will allow the one you are guiding to make the decision. If you are giving interference, you will try to see, or try to make them accept what you wish. When you give interference, you are not complying with natural laws and natural feelings and attitudes, and you, therefore, accept upon yourself, karma. Then do relax and give guidance.

Much guidance will be given to you to put this out into the

world so that these teachings may be given, but interference is not desired. Therefore, do meditate upon this so that understanding may be had.

You have spoken of leaving this Earth before we are ready. Are you talking about the vibrations of the people?

No, not of the people, but of the soul entity.

Yes, of the soul entity. When we do leave this Earth without the proper vibrations of our soul, what are some of the consequences? Will we be around people of lower vibrations, or around people of higher vibrations, and we are not going to be able to understand and exist in harmony with them?

First, you will probably not be around people.

O.K., I'll drop the people. I'm talking about other entities. We will be around other entities won't we?

There are many other forms and many other types of existences, some are with others of your kind. Some are with others of only other kinds. Some are with others and some are alone. Some you will be alone, yet be with many. This is perhaps difficult to understand but when began, and when this came into being as an experiment, the souls who went into your dimension brought with them many remembrances and understandings at that time from other places and lives. The total existences and understandings and remembrances were brought into your atmospheric conditions and were brought in as a whole.

It was not changed as it is now as being born, and with different consciousness existing in that realm, but was brought in as you would move from one city to another. They simply moved into another place and readjusted themselves to be accepted into the dimensional effects that were there. Now, as bringing those remembrances and those thoughts in with them, they were able to continue on and to bring forth many thoughts and ideas and put them all together that made the lifestyles and the ways of existing at that period of time. So many, many ideas were brought from many places. That is

Development is Great on Earth

why there is such a wide variety and so many things on Earth. You have a very large variety of ways to help you to expand yourself, to communicate, to bring into being what you desire.

Many other places are not like this. Many places are of one, two or possibly five different things that you may do or use to communicate until you learn to experience that one thing or those five things together. If it is one, you may experience that over, and over and over, sometimes in different forms, but still the basic concept, until that one experience is thoroughly learned at that point. Then you can move on to another place. Maybe you have, like I said, two to five things to experience there. You learn, then, how to intermingle those experiences and how to handle them all together to make the whole total come out correctly.

You have many, many things on your Earth there. It is a vast variety of experiences and understandings that must all be brought together, which makes it simpler for you to learn. But, at the same time, it can also distract you and can make it more difficult if you fight it. If you do not have proper understanding or proper guidance, then you fight it and you do not get the learning as quickly.

Meditation, relaxation, concentration — these are but part of the things that you may learn to help this existence be easier for the learnings and teachings to come in and flow through you. Many are there within your stratosphere, within the ether that surrounds you at all times. But you have forgotten, because perhaps there is so much variety, how to accept those into your aura and into your body and into your conscious being. These things must be taught to you again — of how to allow the universal knowledge and the energies to flow through you.

There, within your body, has been set up the way these energies and thought forms are taken into your aura and on into your body and then can be utilized by yourself, but this has been forgotten. This is why meditation is being taught there more quickly. The centers of the body that are used to make these adjustments have all but been forgotten. The uses and knowledges that are there now are vast, but are only a very small understanding of how this completely works. Not only from the body, but into the aura and how it is regulated.

As time goes on, I will be giving more understandings of

this but this is why I tell you that you must learn all you can on your plane, because you have such a large variety and such rapid advancement can be made. Where on your time plane say, for example, one thousand lives are lived, on another existence, where perhaps you have only one or two understandings that are basically brought out and taught, then you may only take three hundred and fifty times to learn these one or two understandings. But on your plane where so many understandings and so much learning is done at one time, a thousand lifetimes is not so many.

You have much to bring in together, much chance of when you understand how to bring them together, how to work them together and how to flow into this universal understanding, they may all come through you and bring your vibratory patterns in tune. The advancement is so fast and so rapid, at that point, that your advancements will be very great on your Earth.

That is why I say I do not understand — all wanted to come and now all want to leave. Much learning and much advancement is made where you are now. Do not be so concerned with leaving until it is time to properly leave with what you can properly learn. Much simpler and faster advancement will be learned if you do it all together than to do each individually and then have to learn how to put them all together. Is that understood or is there more needed?

I think that is pretty well understood. I don't intend to leave until all my teaching is completed.

Is there another question?

When the soul leaves this plane, will it go to another level for learning and refinement?

When the soul leaves your Earth pattern that you are now in, it will not jump from there into another universal plane of existence. It will come gradually, as all progress and all development is made. There is never a jump from one dimension to another. There is always a progression of learning and understanding. When you leave the dimensional effect, you will work with and around those who are where you are now.

These will be different types of entities than are now being contacted with clairvoyance (guides). This will be on a more — perhaps Angel types would be a word that would come close for your understanding. Do you understand an Angel? So, although you will be giving guidance, you will be on more of a plane that you think of as an Angel. On a higher soul development progress. Not an Angel, but what you now think an Angel is.

In simple terms, you may move up to where the Angel is now?

Into a dimension of where an Angel would come from, which is not the same as an entity or a guide, but is on a different dimension. Not a different level, but a different dimension.

I used a wrong terminology because I didn't understand it. They are on a plane?

Planes are within your same dimensional level as you may be on now. Dimension is an entirely separate, completely different vibratory effect. You will go into the dimension, and from that dimension there will also be more guidance and changes given to those who are guiding and trying to establish more of a total communication. As I said, you do not jump from one to another, you grow from one to another. Is that understood thoroughly?

I understand it now. I used it incorrectly because I didn't know.

Then, are there other questions?

If your vibrations are not proper you can't move from one place to another until they are correct, is that right?

That is correct, and this is another reason why the vibration must be affected. It is not to make you a good person. It is not to make you a better person than anyone else. These things are for your growth and for your attainment only.

Well, I know we are going to a higher plane, but after we have gone through the lower planes, does it become an easier thing, or does it just continue on?

Learning and working and playing, as has been said before, are throughout every dimension, every plane and every universe. This is a growth pattern, this is a pattern of existence that is carried forth even from the tiniest, the most infinitesimal that you could ever begin to imagine and to the largest.

All planes, all existences, everything works with a basic structure. Such as you understand, that you are realizing that the atom, which you consider so small is the same as the Universe in which you live. There are certain basic patterns that are used and are available for everyone. Learning to work within these basic patterns and Universal Laws is the very thing that you are doing, and we all are doing on every level, on every plane of existence.

When the souls entered here, did all souls enter from the same Universe, or from the same place, or from the same creator?

Souls were together from different places. As all the people walking down the street at one time are from all different cities — from some basic type of soul progression, at that one time, that were gathered together.

Was that like a choice, where they decided to go there for learning, or was it just something that happened?

They were in another dimension. This was from one dimension into another dimension. They were gathered together, they were continuing upon their progression at that time in that dimension. Those who chose, came on into your dimension, those who did not wish to go, did not go. This was not forced. This was a thing that was done from curiosity and from adventure of spirit and wanting new challenges.

The vibrations of the Earth are being raised. Does this have anything to do with the changing of the Earth, like earthquakes?

As to the path of the Earth — the Earth is a living, growing being as you are. Vibrations of the Earth, as it lives, breathes

84

and moves, will continue its own progress. At this time period that is coming up, the time period that has been prophesied by many throughout the times as you request, this will be as a time of growth, spiritually for this entity, whose growth pattern is different.

It is not the same as you, as you are not the same as vegetation, nor of any other living thing. Each thing has its own particular pattern for growth, as for living. This is a time for the Earth to increase knowledge, to increase understanding, that is all. Those who have taken on the karma of the Earth then have agreed to accept and to live within the Earth's karma.

I would like to ask when a soul comes back to this plane, does it select a family where it can experience and it could be the best experience for all concerned?

That is correct, but not always. Basically that is correct. Many times, though, the soul only uses the family to enter into the world. It is not concerned with growth with those people, therefore, you find many times, particularly when one soul in a family is called a black sheep, or that type of person who does not seem to belong to the rest, that may be because he does not belong to the rest. And it is not necessary to belong to the rest for its own soul progression, and may even be hindered by the rest for its own soul progression, and, therefore, there are not the ties and feelings that normally are with a family.

Always, it is not for the betterment of all, or for even the progression of the one, that this happens. But basically, for a basic family unit, this is a blending and a growth together. And not of just one lifetime. There may have been many different lifetimes of all these people who have had connections together at different periods of time that tie together into this one time. But this whole group now may not be this same whole group together in another, and another, and another lifetime.

That confirms my understanding, as my learning is growing. I would like to ask, do different races reflect our vibrations in any way?

Do the individual races reflect vibrations?

Yes, does it mean that certain people return in certain nationalities because their vibrations are higher or maybe they have learned all the things of one race?

In the beginning different areas were chosen and different people of different types were placed in each area because all that were put into bodies did not accept the vibrations the same. Different experiments that had been made were also taken into consideration to bring about different types of bodies and different features. Help was given at that time to bring about the desired effects of the body as had been seen and utilized in different worlds. As the refinements were being made on the animalistic body and the entities were coming into the bodies, the refinements did not take effect correctly, as was desired at that time. Therefore, there were many different developments that were being made then, and different conscious levels were coming about when these things were put together. Therefore, when these people were put in different places, the similar types of bodies, vibratory levels and thought patterns that had been started, of each of these different ones, would be similar. Many times the beginnings were not totally identical in the groups and as some developed more quickly and went away from the original group, there were even many sub-divisions that were formed.

Yes, the vibrations were different and the vibrations have continued to be different, although being raised and changed while much advancement in technological ways, and much advancement in the mind and the body was made through the periods and eons of time. Also, vibrations have been changed even from what they were in the beginning. Many incarnations in the different vibrations, or races as it has been named, have been taken by each individual soul through a normal pattern of incarnations, several times in many of the different vibratory levels, depending upon the desire and the need of that time and that lifetime.

The learning that was desired and the vibratory effect that was needed, to bring the learning about, was then utilized. Not necessarily has each soul developed through each individual race, nor is that necessary to receive the teachings that are given from that particular race. As the time continues and vibrations are being raised, it is now seen at this particular point in the development of these species, there are many that are coming back to be intermingled and become a type of one-development, as it was in the beginning. In the beginning, the separations were desired and were of great help to all, but now the vibrations of all have been changed and affected, and are being changed and affected again, by the intermarrying and the interbreeding of the different racial species and will be continued until the vibrations of all are raised into a unified type of a one-kind-of-race.

It will not be necessary to have the same desires or developments that once were, as that period for growth is over with. No, the development that each individual race could give to the soul at that time had to be learned at that particular time. You cannot go back and learn what you should have learned then, this is not possible, this is not desired. You needed to learn all then, at that time, from that desired group or that desired time period. Things are moving and changing. Time period learnings now are much different and, therefore, the old ways are not needed.

Then the intermarriage of the races is good at this time?

This is a point of progress at this time, the separations are not so desired or necessary now, as was needed for the early time period. All are put together into one. Knowledge, breeding, crossbreeding, vibrations — all is made for the entities to do what is needed to be done at this particular time. All things genetic, physical and spiritual are being taken care of at one time. This is not singled out as an individual type of development on one level. Another question?

When we cease to exist on this Earth (after we die), does the animal soul that we are merged with have a place to go to, or does it go to the same place that our soul goes?

This becomes more technical as we go on with the questions, which is desired. New thinking must be utilized here, more technical thoughts from the different points of understanding will need to be realized. Total understanding by you will be when I give different teachings. Very simply put at this time, the soul, the animal soul, when the agreements were made, was agreed to be put aside although still there, but in the background. The animal soul and the intelligent soul that came into being, these terminologies are used to define for you, is that understood?

Yes.

Alright then. When the two combined, they merged into a third, each having its own ways but still each combining to incorporate into the third, which is man — spoken of as man and woman or the human race. This is very basic, while not totally true or totally being able to be put into words, that you may understand at this time, that still is the basic part. So, I will speak to you of the animalistic portion and of the soul intelligence portion so that the separations may be made on your level. Also understand that this has made a new merging and a new total combination person, but still there are tendencies of one to another, that makes up this new soul development.

Then when this experience is totally over, in a sense, we will be a new soul development?

As stated previously, in male and female experiences on your Earth plane, when one comes into your Earth plane the new development or the new condition must be met and utilized so that the soul development can continue in the pattern there. When the total development for the one experience is completed and you have all that you will learn there, then you will go back to the basic form, although, the understanding and the knowledge and new ways will also be taken to make a more complete, a more total you.

Then we will not only have the understanding of the entity that we were upon entering the Earth, but we will also have the understanding of the animal soul?

That is correct, of course. This is why when coming into the Earth plane these remembrances were brought in fully from other places of being, from other experiences and from other forms that had been taken, and all these experiences were brought in there and together for knowledge. When the soul being leaves again, this knowledge will also be taken with it and added to it, to make the new total development.

Then that's part of the reason why it's so important to refine the animalistic part of our body, because that knowledge will be taken with us?

What you say is proportionately true. More understandings of vibrations and bodies of the two which make the third. Perhaps this would help. The conscious level, which would be considered more animalistic, as said before, may be considered the way of the Son, as of the Father and the Son. The portion that was brought in as a more intellectual soul development that came there, may be considered more as the Father, and the joining of the two, and the development of the two, and the understanding that is merged, and the understanding that will be developing, will more be called the Holy Spirit — as the trinity has been understood for a very long period of time.

This is not as of God that is worshiped there. This is the trinity of the three that makes you, the God portion. So that it is the Son, the Father and the Holy Spirit. That simply means the conscious, the sub-conscious and the super-conscious. The animalistic, the intellectual, and the blending of the two.

And while those are the three main bodies, there are many bodies that also make up these three. It is not as one individual thing makes three. Of the animalistic there will be many parts, of the the intellectual soul development there will be many parts and of the blending of the two, then takes the two and blends them together, but makes up more. This will be explained more as time goes on, so that I may give to you, portions, so that you may understand them. But as a simple basic part, is this understood? Or do you need more clarification?

I believe that it's pretty well understood. You are saying that there are three parts and each part has more.

That is correct.

Would you please give me more about the new developments and when they were developing the new body?

Many refinements were made of this animalistic body that was taken to be made into what was desired for the human development. Thought forms were projected into the animalistic body as it was desired to house the new soul entity. Then, when the necessary adjustments were being utilized and experimented with and brought into a new development, they used the remembrances that they had from other forms of similar nature, for there are other similar forms of bodies in these different universes, which have proven to be in a dimensional level which you are in, very satisfactory. Help was then given, from the dimensional level that the souls were working to get into, and many "Others" were responsible for the final types of developments which were of great assistance to the soul developments. No, this is not understood.

> NOTE TO THE READER: We now under-
> stand that the "Others" referred to are
> the space people and "The dimensional
> level that the souls were working to
> get into" means the third-dimension.

At this time there are many reports on your planet called UFO sightings. In the early periods of your Earth period, there were those who were able to be contacted by those who had been set up to guide and to watch this experimental process and to be sure that there was not harm done. They contacted those in a similar type of dimensional effect (third-dimension) and guided them to your planet so that there may be guidance and help given. There are many remains of these peoples, and peoples — yes, that is an acceptable terminology from your standpoint — and there are many remembrances on your Earth plane of these, and there are many different types that did come. Not just one, but many similar types that did come to offer their assistance and help on your dimension. Therefore, there has been much guidance to the human race, not only the different developments from the thoughts of the intelligent souls that came, but also from the "Other" people who were from other planet type places, that were guided there. Much has been known and understood, although not definitely proven to all.

We have many writings on this where they did come in. Were they from our solar system or outside our solar system, or both?

Much help was given during the early portion of those who came from a totally different solar system, and then much help was given from other universes and dimensions also. From there was developed many levels of beings for help — different dimensional helps and guidances — and there have been many new types of governing factors set up through this time. This is very simply put until more understanding is given to you. But you do know of the progression of people to Angel types of existences, and on up to hierarchies where those who are graduating, so to speak, go on and still give their help back there, until the complete lessons are learned from your dimensional effect, the portion that is called God.

You said, at one time, we have three dimensions here that are all interwound with one another.

This is your dimensional effect which is correctly stated as the third-dimension there. That is correct for what is known. Planets within the solar system that you see, that you know are there, that are revolving around your Sun, are all third-dimensional planets. There are other dimensional planets that are also within your solar system, these are just not upon the vibratory level that is seen or known by you, therefore, they are not a third-dimensional planet. You could call these tenth-dimensional planets or any other dimensional planets that you wish, but they are all of the same type. Therefore, a third-dimensional planet is quite an acceptable name for your Earth.

Then they are actually there, but we can't see them? Our vibrations are of such that we can't see them?

Your vibrations are of the third-dimensional planet that you are on, therefore, that is what is seen and known and understood at this time.

Where does God fit into all of this?

This has been given. God is the portion, the teacher that is for the learning of your solar system, galaxy.

I guess what I want to know is, He isn't within our third-dimension is He? In other words, we actually leave this dimension at the time of death — our souls leave this dimension don't they?

Your souls do leave when you die. Is this the question? When you go from your existence into an existence with no bodies, before you reincarnate, this is called die, death? When you have death you are in another sub-dimension of the one you are in. Dimension is a word that means one vibratory effect. There are different sub-dimensions within one dimension. Within the dimension you are now on, there are other sub-dimensions. There are then larger dimensions, other totally different dimensional effects. Angels would be a different vibratory dimensional effect than death. Death is still within your own dimension, within your own effect. But an Angel is in a dimension that is completely different. It is strange, I understand, the way there are so many things that are the same, but yet are different.

As time goes along I will probably understand it better. I'm not truly understanding your Angel term. How are you using it?

This was just used on a voice tracing recently, but was only used as an example, and this is only used as an example again. Say a dimension is within a sphere, a ball. Within that ball there may be several different divisions — those would be sub-dimensions. Now there may be another ball, a different kind of ball that encompasses the first ball and that is called a different dimension. Terminology will be given. There will be different terminology given for you that may be acceptable into words that you can understand. This will be considered.

I was going to ask you what the purpose of the pyramid was in the beginning. Was that developed by the space people coming in, to show us anything in particular?

That is correct.

Could you tell me, at this time, what the pyramids were really for?

The pyramids were begun and developed by very, very early people (space people who went there to help) to try to help themselves to adjust to the different vibrations that were there and to the different elements they were not particularly used to. As "Others" came to give more assistance, the final type of pyramids, which I believe you are speaking, which are now in Egypt, these were made also to help for what would be stated as a multifaceted thing.

They were used to give continuance to those who were there giving help and aid for this early planet and for information. This was also made and used, to give the people there much information concerning their own bodies and their own vibrations. They were also constructed so as to give to the people of the future, which was assumed would be of more intellectual development later, which is a normal type of development everywhere, that which could be utilized as signals and giving understandings of higher learning and knowledge that higher beings were there. Much as now you send out satellites into space with tracings on them of your people and Earth, hoping that intelligent people or life forms will find them and understand and know that there are other planets there in space somewhere that do have intelligent life forms on them.

There was also a point where the different placements of these pyramids also guided and aided in the energies upon the Earth's development, as much understanding was made and had by those who were there. On all different levels help was given to your planet.

We've lost a lot of it.

That may be true, but still much was given by the "Others" and much development and understanding was given, and was very quickly utilized there in many ways. Many people from La Mureah were aided, and the people of Atlantis

93

were aided even more, and much development was given into Egypt. This was also for those people to learn to develop their own psychic abilities and own vibratory levels. The vibrations were utilized in many ways at that time to help instrumentation on the crafts of the "Others" that were used.

This has been thought of and has been written about here. There have been some people who theorized about this.

Many people do not theorize. Many people give what they know and see, and then they develop a theory, but it is not only theory — it is also of remembrances of their own. These remembrances are had by many who are there now. Many were in connection with all this at that time period and these remembrances are being brought forward as these knowledges are continually being developed there.

Much knowledge will be given through the understanding of the pyramids, but some of this will have to wait. The people who were there so long ago were not developed enough to understand, even had they been able to remember. The technology was not there, but technology is coming very quickly now. As you develop more into space and electronic things — and more sun power will also be of great value to you, it will be seen in the pyramids and will be utilized and you will say, "Oh, we understand this now," as many things now, such as your lazer, and so forth, being developed, you say, "Oh, I understand now what this meant in an olden time, these pictures now represent this," or "This story now tells of this development that we have now." Now, this is acceptable and much will still be done into the future and much will be learned from the pyramids. And, when so much knowledge that you gather will be to a point where you understand this knowledge, then the understanding is brought forth from the pyramids and the leap will be great.

Will any of that understanding be brought through us later on, in other lives?

Other lives, yes.

Are the pyramids still active now?

94

They are always active.

Always been active? But we always thought —

Because they sit there by themselves and have not been used, does not mean that they are not in use.

Are there secret rooms in the pyramids?

There are rooms within the pyramids, but it is not necessary to utilize the knowledges of the pyramids for rooms.

We know of the chambers in the pyramids. Are there others that we are not aware of?

That is correct, there are. There is much that has not been found there yet.

TRANSITION TIME

We discussed entities having free will to come to Earth regardless of vibrations. I would like to know what you meant by entities being sent away.

The choice that all make within the Universes — this is your choice upon the level that you may go upon, that is understood by you now. However, when it was stated that the vibrations would be raised on the Earth and some would not be sent there or they would be sent away, whatever, being sent, this is what was meant: there will be much confusion by souls, when souls try to re-enter into the Earth's atmosphere, and plane of existence, to continue what they wish to do. It will not be possible for them to do so. Much confusion will be had at that time. They will not understand — the understanding will not be there. Therefore, there have been plans made that there will be those who will come to guide and to send those on to a place that will be of either good choice for them or where they may make their own choice. This will be up to the individual soul whether help will be given or whether they will make their own decisions. This will be a time of much confusion to many souls and new, totally new thoughts will have to be made.

You see, normally when you are given this choice, it is at a time when your soul development can see and can understand the progress that has been made and now wishes to go on — to continue more learning or more advancement. And, at that time, the choices that are given are what are desirable for the soul to follow, but this will not be as a normal thing. This will be a time of much confusion and of not the normal type of understanding, therefore, being sent will be a correct term.

In Earth time how long, roughly, before this will begin to take place?

This has already begun. Many souls are not coming in now that desire to. Many family patterns will be broken. The time for learning together, in many instances, is over. This is why it is stated that it is very necessary to learn all you can each lifetime. As explained, you cannot go back to ancient Egypt to learn what was presented to you then, and it is the same now on this development. Many times all things are basically the same, the same types of things go on, on all levels. You cannot expect for your own body to go back now and do what it was required to do many years ago, and all things are progression.

Do the best you can each lifetime with each individual thing. Do not think, "I will do better next time," or "I will make up for this another time," or "I will learn more in my next lifetime." The chances will not be the same. The learning that you will learn in your next lifetime, even though the basic learning will be there, it will be presented in other ways to be learned. So, if you learn all you can each time, then the progression is so much faster and has so much more meaning. This is why the understanding of what you are doing helps you learn so much faster.

Many souls that are coming into this existence now, and have been for some time, are already of higher vibrations than were a few generations ago. Only those higher souls, who have a higher, more refined type of vibrations, are being admitted. So, as those who die from this time and wish to come back from now on, if their vibrations are not more of a refined nature, they will find it more difficult.

Some will still be able to come in that are on a borderline effect that can, so to speak, slip in, just barely slip through;

and if the progression in time is not made — much progression will have to be made in some — they will not be allowed in again. And not allowed — as someone keeping them from coming in — no, this is not correct, but allowed in because of their own vibratory frequencies, their own vibratory patterns that will be matching into the Earth's atmosphere. Patterns, in continuations, must all be as complete each time with each lifetime as at all possible, because it is possible that they will not be there again. Not all — do not misunderstand that. But some may, and you cannot judge who will and who will not be. This is not a thing that can be judged by you.

That's what I was going to ask you, if we will ever know by looking or any method of knowing?

There is no method by which you can physically look or see or tell. This is a thing of the soul and a thing that is from another side.

Could you give us some additional information on raising our vibrations?

Look to all refinement. All things that are of a vibratory refinement. Do not say, "This is not important," or "This is a thing we have always done and is of small importance," or "There are more important things to do." Each thing is important within itself.

Look to more refinement in your physical beings, look to more refinement in your thoughts, in your feelings, in anything that carries a vibration. I have stated many times of anger. I have stated many times of not eating meat and flesh. This is a great refinement. There are refinements that are even considered for only those who are seeking to give total development to themselves in one lifetime and impose much discipline upon themselves. This is for purpose.

Every discipline that is given to yourself is with purpose and is for refinement of the soul bodies and of the finer vibrations. Even disciplines of silence, sex and body functions. All of these things are of great importance to you, although, many people are not at a point where they may utilize the greatest discipline, the greatest thing that you wish to refine yourselves with. But, all refinements must be taken into consideration.

Also, another development, another refinement that may be taken into consideration, because of the strong vibratory effect, is of music. Music and color are of vibratory effects that need to be used wisely and constructively.

Is there a way of knowing the type of music?

All music. Music is a part of creation.

It doesn't matter then if it is calming, quiet music or —

Of course it matters.

The reason I ask is that we have a rock music and —

Do not condemn rock music.

I'm not, I just was wondering if it was all included?

Music of many various types have been given for development on your planet. Many different varied types of development. Rock music is not looked upon as being good, and the vibrations that are put forth into the physical body, on this type of music, does not very often help the physical body level. There can be much detriment done to the misdirection of this music. All music is for vibratory advancement or to create. Vibration is of creation but is often misdirected.

Much thought must be given to what music, and to what level of music. Each person's· vibrations are different. The vibrations needed for refinement can just as well be rock music, as the more soothing music, although this is not the normal. This depends upon the individual, as all vibration is creation, however, it can also be destruction.

The individual must think upon himself in terms of this vibration and in terms of his body vibrations, which will not be the same as his soul development vibrations.

Could this be that the music that is soothing to you, is for you?

That is a good way to think of it. But, do also take into consideration the destructive part of the body. The reason that the more harmonious type of music is considered good is because that it has the vibratory effect within the body of calming or more growing than of detriment. But any type of music can be used to detriment.

Each thing is a thing within itself and can be used for good and for bad, for destruction or for building. Do not condemn any, but all must be looked at in the terms that have been stated. This is a thing upon the vibrations of your world to condemn it, is it not? Much of that is condemned?

Oh, yes, there's a lot of different music condemned.

This is a decision for all, as each progress within themselves. Many need this.

Going back to the beginning. Before we took over the animal body was it a meat eater, or did that happen after we took over the body?

The part that was taken to be refined did not eat the flesh of other animals.

Can you tell us what the body looked like? We have all thought about the evolution of man.

Much is known and much is correct as for the crude beginnings. It is not as thought of as an ape or gorilla. But is of that same branch of evolution where much refinement was able to be made after the help was given. When actually coming down into the final bodies, the people of La Mureah were much as the people of today. Refinements had been made so quickly and they knew what they wanted to begin with.

The others who went out were more the type of animal that you now understand through evolution. This would be a basic type, although, many different ones were being experimented with and developed and were not just sent away and left. Things were not what was desired, so those who came to help the ones on Earth continued to help them, so they were still being worked on and refined even after they were sent to other places.

Through the times these are all becoming very similar — many similarities and developments. As the intermarriage and interbreeding and making of a more one race will be more refined and more acceptable even still.

I'd like to review the way I understand, in the beginning as has been given in the tapes. I understand better now, and even today I understand more.

Knowledge is given in small portions as can be understood each time. Each time more is told and more refinement is given. This is as I prefer to work on this level with you —with all concerned on your level.

I understand that when the souls were entering the physical body to experience, some adjustments had to be made and agreements were made —

Many adjustments were made, yes.

When the souls became permanent in the physical animal body, would this be the beginning of having a soul on Earth?

Yes, that would, for man. That is what made man, human man, yes. At the level when the intellectual soul came into being he became an intellectual person, but depending upon the amount of remembrance that was brought through with him, at that time into that body, was his intellect begun. Those who had no remembrances of what they were before were at a very low level of intelligence, and it took a very long period of time to bring that intelligence up.

The ultimate degree at that time were those who were in La Mureah and was of the highest intelligence at that time. They could use the knowledge that they had brought with them, which was then also incorporated with the knowledge from the animal. So, there were many degrees of intelligence from a very high intelligence to the lowest form of intelligence, which was totally almost as an animal only.

Could this have been at a time when the Divine consciousness or God-spirit was working with the soul?

That is the Divine consciousness. The soul, the soul intelligence that was brought through into the world at that time was, and is, the Divine consciousness. It is the spark of consciousness that is considered the Divine. The animal level is still an animal level as far as what is known as Divine consciousness or a higher intelligence on your planet. This is the only entrance for the type of intelligence that you speak of. Although, animals and other forms of beings have their own intelligences.

As I understand it we are to receive just compensation for our karma. Will this be on a soul level only?

Of course not. This will be on all levels.

Then karma will be assessed on our physical, during our physical incarnation? Depending on how we get in step with the Universal Laws as far as health matters and so on?

That is correct. This is of great importance to you as your animal type body must remain healthy in order for the finer vibrations to be utilized and maintained correctly. Those who are not in good physical health cannot maintain good vibratory levels on all aspects. While the soul development is separate it is also dependent upon the vibrations of the body, so maintenance of good vibrations of the body is important.

Much karma can be released through illness of the body and much growth progress can be learned. Still the soul development is dependent upon the finer vibrations of the different types of bodies that it is surrounded by and encompasses. All work together to make the whole. That does not mean that someone who is ill is a low soul development — no, that is not correct at all — they may be a very high soul development, but still the health of the body will reflect in the other, and will help to maintain the total soul development, animalistic and intelligence.

Then our physical well being will be paid for or rewarded on this plane through sickness or good health? In other words we bring ill health on ourselves by violating the health laws?

Of course that is true. Not only does it go from one incarnation to another on a soul level, but also on a physical level. If you have cancer and are very ill, if you just allow yourself to continue to go down until you die, that is not progression. You may die of cancer, but if you are striving to change that condition, then that will be of physical help in another physical incarnation. The fact that you give up and just allow things to happen to you does not make you in control of yourself.

It has been stated that you are in total control of your own self and your own energies. You are the one who creates what happens to you, now or any other time. Progression that is made through an illness must be made. You must not just give up. Then nothing is gained. While help may be given on the soul development level, if you give up on the physical level, you will not have any physical growth and this is of great importance.

In other words, that matter of health would be the same as soul progression, strictly a personal thing through your own efforts?

Absolutely. Can you expect another to give you good health?

No.

Can you expect you to give yourself good health without the proper mental and nutritional guides being followed?

No.

Of course you cannot; that is only common sense. People must understand that these are part of the refinements that are being told of and are desired to continue, because this soul, that is developing, is still a part of the whole. The vibratory effects that were set up in that physical development are taken into that other portion when it is not within a physical development. It will then continue into the next physical incarnation.

Also, when this portion, that is called God, is obtained and you do go on, out of this learning period into another, then all of that must be settled. It must all be settled and completed before you are allowed to go from this portion, that is called God, into your other developmental stages. So, all of these things must be met on the soul level and also the physical level. New teachings and new developments will be given to you there.

There are times when a soul is so developed, but there is some small portion that is left, that it may go on and learn that lesson elsewhere, but this is not a normal thing that is done, so do not expect to be one of those few that are taken into different consideration. They are very few.

When a person on this plane refuses to take those necessary steps or discipline himself in any manner, does this indicate that he is on a lower vibratory level or just doesn't care for himself?

This question can be answered different ways, from different levels. This is a difficult question to answer with just one answer, from one level. Simply, it is saying that this person does not care for their body, this person is not developing properly in their bodily development. It does not mean that the soul itself is not within progression. It means that it is not utilizing its progression to the fullest.

While manifesting illness is one thing, not taking care of that illness is another thing. And, while total lack of help for themselves, total disregard for their body is on another level, it does encompass the whole. This does not mean that their soul is not a good soul. It is just that they are not making the progression that they could be making at this time. You cannot say that because a person has a body that is ill that the soul is not making progression, that is not correct at all, but they are not doing the best they can. This may be a karmic thing that they are learning, striving for, trying to reach. All must strive and utilize everything that they can to advance themselves totally.

When we were talking about in the beginning, you said that some of the souls took bodies too quickly and were damaged and they were sent back for redevelopment. Could you tell me about that now?

This will not be correct, but this will be as correct as possible for your understanding at this time, and later when you are given more information you will say, "This is not what you said on the voice tracing."

The vibratory elements that are within all, and are all, were not met well. The vibrations of the soul development were damaged, for lack of better words and descriptions here. That soul then was not able to continue within the vibratory frequencies that were set up on the Earth dimension. Therefore, they were collectively sent or returned, meaning basically the same, to a place where they could redevelop themselves. And, while this is a very crude comparison, as a person who was an alcoholic would damage himself and would go back to a place where he could change his vibratory patterns, and redevelop in a more constructive manner. This is a type of understanding that I believe is close enough for all to understand.

I think what I was wondering about was, if they went to another place of learning?

Yes, to another place of learning, where they could redevelop the vibratory patterns that had been damaged.

Then their vibratory pattern had been erupted in some way, and through the learning would be corrected?

That is correct. Some of these things are difficult to put into your language and into your understanding until you gain more knowledge. You will understand this more clearly soon.

Thank you, now if you will release "This One" —

This connection is not ready to be broken. There is more. There is one more thing that I wish to say at this time. This is a good time to say this for all concerned, everywhere.

The thought is out within your world, and was stated correctly for that time, that vibrations go out from the body and collect like vibrations and come back and bring the like vibrations back to you. Is that correct?

Yes.

That is correct? That is **not** correct. That is not the correct thing. That does not happen in that way. Vibrations are formed from you constantly. They go around your body and they are projected into the aura and into the finer vibrations. This then puts up a vibratory effect around you and, as a process called osmosis, where you absorb like things into you, is how this works.

So if you are negative and you have negative things to think and to say, and this is the basic type of vibration that you have built up around yourself, when the positiveness comes to the negativity of your vibrations, then the negative consumes the positive as fuel. That is very simply put, but as a basis it is consumed, and it is not allowed to enter through, so you continuously receive negativity, or negative things happen to you.

Most people there have a force of negative and positive alternating and you are able to raise you spirits, etc. However, still the positive can only come through where the positive is, and the negative can only come through where the negative is. So, you must think of refining your vibrations and keeping them on a level where the positiveness is ready to accept more positiveness, if you expect, and if you desire to have good things happen to you, and to keep yourself in good spirits. Is that understood?

Yes, it is.

Anger is of a good thing for a person, because it can clear out the closely knitted anger that builds up within the vibratory areas. However, bursts of anger to clear out this that builds up on the vibrations is not desired, for the built up anger is not desired. It is more desired that you learn to control the vibrations that make anger so that it is not there.

Your thoughts and your words, which are extremely powerful, are all within your aura. When words are spoken of crude or negative vibrations, you are making yourself crude, and then crude vibratory things come in through your vibrations and it just continues on and on. Refinement of the aura needs to be made, words need to be selected to give proper direction, and to create what you desire. Do you desire to have all, to have what you want on your Earth, to be able to create the things that you want, to have an abundant life there? Then you must not only keep your mental attitude good, etc., but you must use the tools at hand that you have to keep the vibrations refined. Refinement of vibrations is extremely important, as said previously on this voice tracing. Are there any questions on this?

No, not at this time.

Then this connection will be broken.

Chapter Six

(Tape 16)

Could you give me more information about the portal that was between the dimensions in the beginning?

The portal was as a window from dimension to dimension. This was just a place that came into being in that dimension where they could see into another dimension.

Was this portal open to all souls in that dimension?

All who wanted to look, could go and look.

I wondered if the souls who first came down were of a higher vibration than the souls who came later?

Souls that went through and were the first that began to experiment with this, were more adventurous souls.

Does being more adventurous have anything to do with vibrations?

It means that they had more adventure within themselves. It has some to do with vibrations, yes.

When the souls first came down, they did not enter the bodies did they?

Not at first, no.

They were just using thought forms? Could you tell me a little more about that part of it?

There was a long period of looking before anyone tried to go through the portal, and several things were tried, several experiments were made before the first did put themselves through and, when this seemed to be all right, then others came through also.

Can you tell me anything about the experiments that were tried?

This would not be understood at this point. Just pretend that you would be trying to see if things would go through and come back and not be harmed.

After we came down and began to enter life forms, there was some dissension among the souls. Please elaborate on that.

When the intelligent souls that came into your dimension decided to try to enter some of the life forms, they began with whatever would accept their vibrations at that time, and were crowding out the other soul developments that were already there. This brought about much trouble and many problems, as it was disruptive to the soul development it went into. It was not allowing the normal pattern to continue.

Then at this time there was a decision made — the agreement made, that only the one body, housing man at this time, would be taken?

That is correct.

Then there was refining done to the body? Was this before we took it or after we had already taken it?

Much refining was done before, much work had to be done to make the body to house the soul entity for what was desired, and also to make arrangements and changes in the body so that it would better accommodate the soul.

Was this done through breeding?

This was mostly worked through mental processes on the glands and on the brain and through the different nervous system processes.

108

Then the brain was actually enlarged from what it was?

Enlarge is not a correct word.

What word would you use?

Electro-circuits were rearranged through mental processes.

The reason why I said enlarged is because the scientists say we have such a large brain and we use such a small portion of it.

You do not utilize what you are capable of.

Yes, but as our vibrations are raised will we not utilize more of this?

It is desired.

Is that desired by us or the Universe?

By all.

At the time that we took the bodies that we inhabited, there seemed to be some mutations. Can you tell me a little more about the mutations? Where did they come from?

Mutations? Of what?

At one time you said that there were mutations. My thoughts would be — something that was born as a mutation, things that were not quite right.

That is acceptable.

Can you tell me more about them?

Of the animal mutations? Yes.

Were we partly the cause of this in the beginning, entering the bodies?

From entering the bodies, also from experiments that were made that went on later. But, also, some of these were from the changes that were made in the beginning in the animals from the soul entities' vibratory patterns mixing with the animals' vibratory patterns improperly. Therefore, in the animals that were born later, the genetic codework was severely disrupted and, in that way, many mutations were made. This also was part of the disagreement.

How was that part solved?

The agreements that the soul entity would only use that one body. This was the quickest way to keep all happy and to give them a way to experience upon the planet and in the dimensional world and also to leave others alone.

I must ask at this time, when we first came here, was it thought of as a place of learning or just as a curiosity?

In the beginning, the Earth was a place of learning for the soul developments that were there. At that time, the coming through from the other dimension put a whole new type of entity upon the Earth. The new thinking that they brought with them did change what was going on there. It interrupted the normal cycle that was being developed. However, it was accepted that this would be.

The ones who came did not come for a definite learning process as they looked over and said "We will go to school there." No, they came over just as you would go somewhere, because it looked interesting and you wanted to see what was there, to experience what that particular experience could give to you. It was, you could say, a learning process but not exactly the type you are thinking of.

Are new experiences like this in the Universe considered to raise your vibrations?

All experience gives new knowledge. They did not come into your dimension to get more universal knowledge or raise their vibrations. This was not the reason.

Do vibrations reach a height where they maintain themselves or are they constantly changing?

Although the point sounds simple enough for you, to give you an explanation at this point would not be accurate, and yet would have to be told accurately to give you the truth. This will be explained to you a little later on as we discuss other world cycles. All things are in cycles. Cycles are part of the universal pattern. Vibratory cycles are also part of the universal pattern and will be explained at a later time as stated.

As our vibrations are raised, will there be other planets that we will be able to see?

If your eyes are changed.

I'm wondering, because it seems like there are new planets that have been coming into view.

As the vibratory changes come about there will be some physical changes also. There will be new things that will be seen, there will be new colors that will be seen, there will be different wavelengths that you will be able to hear also. However, your planet is not quite ready for this and the possibility that other planets will be seen is within the future, but do not expect this immediately. This is not for right now.

In the beginning, were any of the animals that were here, were they meat eaters?

The animal kingdom is different in its evolutionary status, as is the plant kingdom. There are agreements and understandings between the souls, and they are on their own development patterns. This was taken into consideration for your dimension on your world also. Many of the animals, even at that time, did eat flesh, however, their reason for eating flesh is considerably different than man's reason. What is handled by the animal kingdom is to keep the animal kingdom strong, this is known there.

The strongest will survive, yes.

Also, each one has its own things to do to keep your planet in a shape that all may survive. If not, dead, decaying things would ruin your planet quickly. All fit into a pattern and all have their own place in things. This was quite well within their development pattern. This has nothing to do with the human species now. This is a different development entirely, although taken from the animal, it has now turned into a totally different evolvement.

What about animals that we are now putting in parks and zoos? They are not really allowed to run free like they once did, which I'm sure has changed their life patterns.

Of course it has, and while it is very pleasing to the eye and it is very interesting, that all may see these different species, it is not a normal thing for these animals. Much help has been given in certain areas to keep some animals alive. In some ways this is good, but in other ways, seen from other places, I'm not too sure that it is good, as there are cycles to everything, things come and go. However, it depends upon if that one cycle is ready to end yet, if not it will continue on. Therefore, since help is given in that direction, and is considered as help, it will continue to be accepted as help. Is that understood?

Yes. Who made the changes in the animal refinements in the body? Was that done by the "Others"?

The soul developments that came down worked along these lines, and had the basic pattern in mind that they wished to use. While experiments to get the correct pattern that they wished did not always come out as were expected, then the guidance was given. As you see, they were working in a totally different dimension than they were from. This meant that hardly anything was the same, hardly anything worked in your dimension that worked in their dimension. While they had a vast knowledge from where they came from, still it was hard for them to make it work in your dimension, as things did not work the same. Therefore, there were ones guided to Earth to help them that were of the same dimensional level that they were now working in. They were guided to Earth to help them complete the experiments that were necessary for them.

Were the babies always born the same way, or was that altered too?

No, that was not altered.

In the beginning did we have guides?

In the beginning this was not set up as it is now. There were only soul developments that came through. There were only soul developments that were there. There were not the normal guides that are there now; there were not that type of guides, but there were those who watched and who were guides, such as where I come from. But the same type of guidance was not there, as with other entities and guides now.

I understand there were guides from other Universes and from other areas. I was thinking more of our guides that we know today, how they came into being.

They came into being as many of the new creations, and the new avenues of evolvements living on your dimension came into being. As experiments were made, as souls progressed, and as different levels and vibrations were set up, as all these things continued to go along a normal pattern, there became divisions and there became different vibratory levels that some were able to survive on. This now I am speaking of the new soul developments, after the body they chose to enter into ceased to be and did not continue, and they were again set free. However, they were a new and totally different soul development than they were when they came into your dimension. They found the vibratory rate and the vibratory level that they were comfortable on, and many came back into bodies when it was their turn or when it was possible for them to begin to enter as the experiments were made, and they found that they were able to enter through this process known as birth.

Many adjustments had to be made on their side also; many experiments had to be made from their level to get this into being. This was a very long period of experiments and developments that was watched and guided carefully by "Others". But still, there was much experimentation that was going on until this became a regular sequence and all the sequences seemed to fit into a certain pattern that was acceptable within your dimension. Is this clear?

Could we say, then, that some of the souls that entered bodies —that when they left the body, they found another way within the dimension to help some of the others while they were waiting to be reborn?

They were waiting. They did not, at that time, actually help others, but they were working within themselves at that time as there was still much consciousness that was being held in. Not as babies being born, you see, for they came in with total knowledge and they left with much knowledge also. This knowledge still left with them, at that period of time — the first ones especially, and while they were there, although of a different vibration, were still not of the first vibrations where they came through the portal, because when they mingled with the animal, this made a different type of vibration and it made the different type of person, or soul entity, that had been told of before.

They were still of a different vibration than those who were still waiting to go into the animal body, but there was still understanding there and they gave from their dimension the help they could give — still experimenting and still seeing the problems they had while trying to find some solutions for those problems. During that certain period of time you had many who had come through the portal who were still that same soul development. Then you had many that were in the body of the animals and also, as I said, there were many levels and sub-divisions of the animal people that became people, so there were many kinds that were on the Earth at that time.

Those were sent to other areas, you said?

Yes.

114

But I don't quite understand that part of it there. I thought I did, but I see I don't when you said "animal people."

Well, when the souls went into the animal bodies some were successful, and some were not as successful, do you remember that part?

This is where some lowered their vibrations and were having tremendous trouble?

That is correct — some were — the ones that were not successful were put into groups according to the experiments that were made on them and also to the vibrations of the development of the intellect, and so forth, that they had, and were put on different parts of the Earth in groups so that they might continue to develop. At the same time, you have some that the animal body had deceased and the soul development had changed, because it had mingled with an animal, and was then not in the third-dimensional world with the animals, but was on a more vibratory plane as the first soul developments, but still separate, because they now had the animalistic part with them, which made the new entity, the new soul entity.

Which could not go back through the portal?

That is correct.

Then, is that when the decision was made that they would work with us, such as our guides today?

Yes, a little later on — not at that exact time, because those were, you might say, scientists. In more understandable terms, let me explain it this way. You had the scientists of those who came in through the portal, you had the animal soul and the intelligent soul that were combined into the animals. When they died and went on, they became the combined soul entity. So you might say that they were the scientists of the new soul entities, because they knew the problems and could work from their dimensional level also.

So, at that time, there were several that were trying to

help, plus you had those that were sent to Earth from other planets that were also working on the bodies. At that time, there was much help being given and much work going on from many different levels. Eventually, some of those souls saw that there would be guidance needed for the souls in the bodies when the birth process became more perfected. They could see many problems coming about for those who still had the original knowledges being born into tiny babies. So, it was decided that much help could be given if someone from that dimension was there to guide, or to try to guide, and to help along the new entity as much as possible. Within this group there was Mar-che-ah, "This One's" guide. As you know, the name e-ah means "from the original" or "from the beginning." She was one who was within that group period that made this decision.

Linda and I thought that she was one of the first.

Alright now, the ones we speak of now are ones in a new dimension. After the animal was deceased they were on a new dimensional plane, and this is what we now speak of. All that were there wanted to help, or to do something, as their world also had many problems and it also had to be developed. So, as an animal deceased, the soul was then given the task of guiding those still alive, so those were basically the first guides.

The ones that still had total knowledge were then freed to solve the other problems of that dimension. And some never took bodies again so that they would be free to work in that dimension with their total knowledge, and so, Marcheah is one of those. You see, many problems come with developing a new world, a total world, and a totally new development, many problems from many areas. This totally new development had not been basically found before in the Universe of this great magnitude with so many problems, and so many new learnings and new teachings. The magnitude is greater than you could possibly imagine.

That helps me understand guides better. During the beginning, when the vibrations were lowered in some, and they had to be sent to different areas, is this where we got the different races, due to their vibrations changing the bodies?

Not only due to the vibrations, but many experiments that were made, many changes were made, many things were changed without knowledge of what the outcome would be. When certain ones fell into certain group types, the similarities were placed together and even after the similarities were placed together, they were continued to be worked upon by the "Others".

From these basic groups were sub-divisions started. Some went together who liked each other better or had similar ideas, or wanted to do one particular thing and the others didn't. There were many small sub-divisions that came from the basic pattern groups that were originally put together.

You say they were moved. How were they moved?

They were moved by those who were guided to come to Earth from other planets. They were transported and put into a place that seemed to be the best place for each group type at that time.

Then that had to do with the elements at that particular place?

That was part that was considered, yes. They were not groups that were put there and then developed their own certain type of things for that area. They were put there because their certain types of things fit that area.

I'm glad that you made that clear. Then actually we were more all one in the beginning and then divided? But as knowledge progresses and vibrations are raised we should all be of one race again?

While the separate groups were desired at one time, this is now a period of transition. Eventually, being all of one race will be the best for all concerned.

Were we still in La Mureah at that time, when all that was going on?

It was during the time that was before the physical La Mureah, during the dimensional La Mureah, and during the beginning and during the whole thing. These were the things that were going on. It was as you would go to the moon and looked around and there was great excitement, and you found ways to battle whatever the elements were or whatever creatures were there or whatever you had to do, you did it. Then, all of a sudden, you found all of your spacecraft destroyed and you had no way to return back to Earth. Then what do you do? You do the best you can, you make out, you accept the fact that this will become your new home and you begin to cultivate it and live on it, and find the proper ways to develop yourselves there on the Moon. This is much similar to what happened to the souls that came from the other dimension. Although in a different way, but still it is the same type of thing that happened.

Were there other souls who came in, besides the ones that came in through the portal?

Not at that time.

I'm talking about later, are all souls that are on Earth today ones that came through the portal?

No.

Did all guides come through the portal, such as "This One"?

"This One" did not come through the portal.

I'm talking about the guides that were sent, that had the connection with you and your Universe. They did not come through the portal?

No, they did not.

Now, not counting them, did all the rest of us that are here now, did we come through the portal?

Since that time began, and since those original souls did come through in that way, and seemed to be trapped on the Earth's dimension, there have been some that have gone through into the Earth's dimensional effect that did not go through the portal. No, all that are on your Earth did not go in that way, but there have been many alterations in vibrations and many things to allow this experience to take place there.

Does that mean that those particular ones would be able to go back where they came from? That their vibrations were not lowered to the point that they couldn't go back, such as the ones that came through the portal?

This is not the same. You are individualizing again. Ones that did not go through the portal, such as "This One", depending upon the total vibratory effect, reason and so forth, of entering into that experience, those things would be taken into consideration.

In other words, she couldn't go back, right now, until her vibrations have been brought back up?

"This One" can go back to where she was, when she pleases.

I was wondering if there were souls coming through from other Universes that were able to take bodies to help, or whatever they do, and after their body ceases to exist, go back to the other Universe where they were?

This can be done.

Well, I didn't know. I guess who I had in mind was Jesus. I felt that he may have done this, I'm not sure.

No, this is not correct.

You mean he did come through the portal originally?

He was one who did come through the portal, and had many lives on your Earth before he became your Jesus.

Is there anything else down these lines that you would like to say?

One thing that I do wish to state, that seems to be as

many things are there, a misconception. While true to a small degree, it has been made into something that is not totally correct and the proportions have been distorted severely. But much is told there of the souls that came in through the portal and, for sexual reasons, this all occurred. This was not true. This was one thing, and not even a big thing. It was, as to say, that they wanted to smell the beautiful flowers, and then blow that out of proportion. That the only reason souls came was to smell the flowers, this is not correct. The basic reasons have been given and the sex was part of the curiosity, but it was not the damnation of the soul. This was only part of what happened.

The mutations were not basically caused from that thing. Mainly, this was from distortions of the vibratory fields that were intermingled that changed the patterns of the growth and the patterns of the cells themselves. Some were because of intermingling of sex and the "Others", who came to help in your dimension, did find a way that this was stopped. Once the mutations began they were passed on from generation to generation and they continued to multiply rapidly. And this was the reason for so many mutations, not only physical but mental, many mental mutations from both animalistic and from the humanistic were formed through experimentations that went on during that period.

These mutations had to be controlled, they could not be allowed to go on uncontrolled. Some controlled these mutations mentally to allow the others to continue to live and to experiment and to find better ways. The "Others" were working with the animals and the mutations genetically, so to speak, to try to get this stopped, and when the time came when this was more under control, at a much, much later date, then there were those who were given the knowledges of how to help those who were born deformed. Many people then worked to try to correct these mutations and to alleviate the people and animals of these deformities that had happened to them. While this was spoken of, it has been made into something much more than it actually should have been.

These teachings should bring new understanding to the people. Are male and female only in this dimension, or is it experienced elsewhere?

In the Universe there are many different sexes. Male and female are just two. There are only two in your dimension. This is sufficient.

But in other dimensions there are more?

There are others. In the beginning there were two different types of bodies to take, then whichever body was taken was what you became. It was not a great decision in the beginning, "Shall I become male, or shall I become female?", because becoming male or becoming female was not totally understood, even then. See? So, at that point, it was not a big decision to be made, was not a big consideration and possibly even now, as large a consideration as people may put on it. Sometimes it is, but not always.

In other words an entity can come back either as a male or a female?

Whatever body is offered to them.

Our Bible tells about Adam and Eve, that they took a rib from Adam to create Eve. Is this just a fable?

This is how it was thought of at one time as man could understand that part. It was a good way to find a division, and also to see that the female was subservient to the male. Many things were entered into the stories then that are now understood differently. Many conceptions during that period of time were entered in the way a story would be told. Stories during the days of the Bible would be told in consideration with those days, and those knowledges. The same story could be told now, but would be put in a more modernistic version.

I am aware of this. In the division of man and woman was the woman always the weaker sex and the man the stronger sex?

This is the way it was with the animal, yes.

You mean the male animal was stronger, the female animal did bear the young and was not as physically strong?

This was the way of the animals. This is not of dominance but of natural law. This was for the animal world and this was for those who were not as successful. For those who were more successful and lived in La Mureah there was more equality, because those who had total knowledge, all had knowledge and understood that taking of the male and female body at that time did not make one more subservient. There was an intellect there, you see, an intellect that made more all equal, and this was part of the La Murean way. While it was considered that the female was not as strong physically as the male, and there was more consideration given to her also, as the way into this world of the new soul developments — these things were all taken into consideration, but on an intelligent basis. They were all intelligent, they were all the same, and all had as much to say as the other for that period. In that one area in La Mureah, that is how it was. But those that were not so successful had to live with their own ways — whatever the ways for their time period and their development or their group decided, and had to live out whatever it was in the way of intelligence at that time.

Did the knowledge and intelligence of the people begin to decline before the earthquakes that destroyed La Mureah?

La Mureah was, in your years, a very long period and many cultural changes and many changes of many types were made. I cannot say that the culture went downhill — it changed. In some ways, things were made better and in some ways things were not made better as new thoughts and new ideas and new understandings of this dimension came about.

You mean changes for what was needed for the time?

That is correct.

Then can you tell me, in our Earth years, how long did La Mureah continue?

This is difficult again, because Earth years are not the same.

Oh, their Earth years were different from ours now?

That is correct, but this was for a very long period, and life span was not the same also.

What was a life span then?

Life span was much longer than it is now. So between that and the difference in the years of then and now, it is not exactly accurate to say, but for a very long period of time. But La Mureah was many, many thousands of your years.

In comparison to our life today, can you give me any kind of life span?

Different parts of La Mureah, which was also subdivided again into different life forms, but also in vibratory fashion of those who first came in, also made a difference in the length of the life and the length of creativity of the cells of the bodies. So there were life spans that would go for many hundreds of years of what your life would be now, many hundreds. This was due to the vibratory effect of the entity that entered the animalistic body.

When the eating of flesh is discontinued and the quality of our food is raised and our vibrations are raised, will we return to living 150-200 years or even more?

What you just said is possible and can happen, but do not expect it to be the same as when they came from one dimension to another. The vibratory effects will not be the same as when those came through, no. They came from a different dimension to your dimension and that was brought with them. Through the generations this has been lost. That is not what you wish to do now, that was for then, it is not for now.

We lost some and gained some.

That is correct.

Did I come through the portal or did I come in another way?

123

You did come through the portal. You were one of the first to go through the portal.

This sounds like me, I'm very adventurous.

This has not changed through your lives. This is still part of you. No matter what cycle that you have come through in your past lives, there still must be the place when you come through that will have this same affect upon your life. This is not understood. I will put it more in terms that you may understand. When you pick a life, and you pick an astrological sign to come through, if it does not suit this curiosity, then you must wait until the other stars are in position that will shine and bring this curiosity into being upon the one that you desire to enter through — through the being that you come in, to take over the body. That body must have the astrological influence to provide that same type and height of curiosity and achievement. This is with every life that you have taken there.

Is that good or bad?

Good or bad, this is your way. This time the sign that you picked has all the things within it to make you, and to help you. But, had you picked or come into a sign that did not have these qualities, then you would have to wait until the other stars come into conjunctions to make this quality in that sign. Is that understood?

This would be my decision?

Of course, this is your decision.

Under what circumstances were Linda and I brought together in Atlanteah?

In the beginning, when this was arranged, it was because of your great and vast knowledge in Atlanteah and because of her special capabilities.

In the last part of La Mureah were the vibrations still being lowered?

Technically yes, that is correct. The people's vibrations, through the generations, were being lowered from when they first came in through the portal. That does not mean that they were becoming bad people. This terminology, this is something else that, when it was understood, was misunderstood. This is not a downfall of those people, that made them bad. This was just the lowering of the vibratory effect, that was brought from the other dimension. Through the generations they were losing the connections with the knowledges that they had. They were losing, you might say, their power. The power to utilize what they had and therefore, the knowledge of what was, or what had been at one time, was brought down. There were some that did hold this through their extended lifetimes and when they did choose to come back, these knowledges and these powers were still with them. Even into Atlanteah and Atlantis many people, such as yourself, still had powers that the common people, or the masses of the people had lost already, or many of the masses never had them.

When you talk about thought-forms, does this mean the materialization of a given object?

Technically and basically that is the definition, yes. The thought can be brought into being — materialized.

Is the power to do this given to them from another area or is that power carried within their being?

That power is carried within the aura.

To materialize?

That is correct. That does not mean that help cannot be given if it is so deemed necessary and deemed correct.

In other words, with the proper knowledge at that time, if they needed extra power to do something spectacular then they can draw from other dimensions too?

If you had this knowledge and still had this within your understanding, the answer would be yes.

As our vibrations raise and we go back into the Universe when our time is over here, will we gain any of that knowledge back?

Much knowledge is being gained now. However, do not forget, these things may not be of use to you where you go next, as many of the knowledges brought through the portal were not of use. Their knowledges were of that dimension and were not of much effect on your dimension. Some they found worked but in different ways, and many worked well, but many did not work at all.

When did the space people come in to help us?

There were some who were there close to, very close to the beginning, not long after the portal. They were guided there to help those who were trying to bring about the changes, and were having a hard time working in that dimension.

Did they come here because of their special knowledges or did they just come to do what they could?

They came in to do both.

When the big earthquakes came and La Mureah was totally devastated did it happen all at once or over a period of years?

Much of La Mureah was at once.

Then Atlanteah sprang up next?

Atlanteah was already begun before La Mureah was sunk.

From things we have covered in our past lives, Atlanteah seemed a little more primitive.

Do not forget that as the vibrations and the powers of the people were beginning to lower genetically, they were sent to the new lands and were trying to develop new areas. The primitiveness of those who were there in Atlanteah was not exactly primitive, as some who left La Mureah took much power and knowledge with them to develop Atlanteah. Atlanteah was begun of a class of people with very high success but not as high as La Mureah.

In our past lives we keep coming across sub-humans. Please tell us more about this.

In Atlantis, there were those who came from other places (space people) to help, as was told earlier. They were working to change the genetic codes of the animals and to stop the inter-breeding that was making more mutants and to find a way to keep this from happening. Also, many experiments were going on to help to remove from the animals, and also from the sub-humans, as they were called at that time, the appendages that did grow from them as part of this genetic change. There were many experimentations, and much knowledge from the "Others" was given to try to help, and a mingling of knowledges and brand new technologies were made then. New machines, of sorts, were made then to try to make corrections on this. From the knowledges that were learned at that time, the one that told this not long ago to your people, by the name of Edgar Cayce, did learn what he learned in Atlantis in that period, and was part of that.

He was helping in that period?

He was. And when the signs were that Atlantis was to be destroyed and that there were problems seen, he did take a band of people and go to Egypteah.

Egypteah?

At that time, later was Egypt. Egypteah was also a place that was of the original that was developed. At one period of Atlantis, "This One" (Linda) and her husband did go there, with help of the "Others" who came in, and were helping the

Egyptian people to develop their lands.

This was the time in Atlantis when the space people sent Linda and her husband to Egypt? You don't mean Edgar Cayce was her husband, do you?

No, but his knowledge came from the time period in Atlantis where there was the experimentation still going on. Many of the sub-humans and the animals that were deformed were sent to Egypteah to get rid of them, so to speak, and there was great need when he went there.

It seems to me like there were a lot of people coming in from outer space to help us, even now there seems to be more activity at certain times than at others.

That is correct.

I feel that "This One" is tired and so we will close for now.

Chapter Seven

(Tape 17)

In studying past knowledges that have been given to Earth, I find a lot of variation.

Many things that are given to the world as a truth, to help bring knowledge through, have been distorted, and are then expanded upon, and become truths, but still are non-truths.

Truths for that time, but not truth for this time?

But they are changed. Something that may be a truth, but is not something that is commonplace, then is taken to be commonplace. Such as one question that has been asked, that has been thought of and talked of, but has not been asked of me yet, is the matter of walk-ins. This has been of much discussion. This is one thing, that while it is a basic truth upon your dimension, it is not a commonplace thing.

This thing is a very difficult process to accomplish. It takes much determination upon the entity that wishes to do this, and mostly cannot be accomplished unless the entity is of very high frequency and can withstand this. Also, the one who's body is given over must have certain vibratory effects that will allow this to happen. The agreement must be made between the two. This is an exception more than a rule. This knowledge was told and now everyone thinks he is a walk-in.

That's true.

This is used as an example. As so many things are told,

this is what happens on your Earth. This is why knowledge is very difficult to give. So do understand that there is such a thing, but it is very rare. This is not commonplace.

What about ghosts? Are what we call ghosts manifestations of beings that refuse to leave our area for some reason?

Most knowledge in this area has been expounded upon, although there are distortions, as stated, but basically what is stated is true.

There are some people that linger on?

Not people, no.

Well, I mean souls.

Not souls, no. These mostly are astral shells, astral bodies that still linger on. For energy is still there or this portion of the body is still there. This will be discussed more as bodies and understanding of bodies and vibrations are explained better.

Have there been more space people sent to Earth to help, as the connections have been broken with the ones that were sent here, like "This One"?

The basic answer to that question is no; however, this is not 100% no. These space people have been sent to Earth since the beginning and this was not the original intent, but this is something that is being worked out through them now. This is, from time to time, a necessity for the two-way communication as desired.

About when did your communication break off with the people from your dimension? Was it La Mureah, Atlanteah, down through that era or later?

This has been gradual and the lowering of the — it is not exactly the lowering of the vibrations of these entities, but it is the keeping of the thoughts and the vibrations at a lower level. They had lost their way to expand their minds and change their frequencies. If you do not reach out to where I am and keep the connection going, then it is lost.

The religious people are still expecting Jesus to return to give the new teachings, and that is the only way they are going to look at it, at this time.

This may have been told, many things were told.

This is what they are waiting for and they don't want to accept anything else, at this point in time, except another Jesus.

Well, since the teachings were not the original teachings of the one called The Christ, how can they wait for more teachings to come from Him? The teachings they follow now are basically not from Him. His teachings have been distorted and changed.

Many on this Earth are publicizing that Jesus will return.

Strange, with so much publicity, even now, more and more are having higher thoughts, different thoughts. Think on that.

Oh, I'm sure, but they are not being shown —

You do not understand. With so much publicity of Jesus and His teachings, there are more people seeing and hearing of this than before, correct?

Yes.

Alright then, look to the large numbers that are free thinkers. There are more and more, are there not?

Yes.

That should tell you something. They are free thinkers and they are expanding their views, and they are expanding their minds, and their thoughts and they are reaching out.

That's true. I'm not knocking them, I'm just saying —

That's better than if there are no free thinkers.

Yes, I know that. We went to a UFO meeting the other day and, listening to the speaker, I felt like I would like to share my knowledge with him, but I didn't feel that it was the right time. There is a lot of misunderstanding in this area.

That is because you know the truth.

I felt that I would like to reach some of the people teaching these groups.

This will be done.

In Jesus' time were there space people coming in?

Of course. Your great literature has many historical accounts of them. Look to your Bible for some.

Maybe we can help to straighten out some of the misunderstandings.

I am not here to correct the world. I am not here to change the world. I am here to give the correct teaching for those who are on a level that may understand. Those who are not on that level, and wish to remain where they are, is perfectly fine with me. But for those who wish more, I am here to present the more they are looking for. This is all.

At one time, you said I was on vacation before I came into this world and I just wondered — were all the souls on vacation before they came here?

Where the souls were was not considered a vacation, no. But there had been much time before, for you, not long before, that had been as a time of rest and recreation for yourself, yes.

Will it ever be possible for the animal souls to make progression into the human body?

This thought has been given and has been explained as progression from one kingdom into the other, as intelligence does advance and as the soul does advance. At the time that this was given and thought of, this new knowledge that was coming through was fairly revolutionary, and was within good dimensional thinking. However, it is not correct. While all have souls, and all are souls, there is a development of what the soul is learning. The soul takes upon himself this stage of development in different times in different places. All have different developments. But it is not as graduating from the mineral to the plant to the animal to the human in the Earth's existence time.

Each thing is considered separate and each development, while similar, is still separate and set aside. Each vibratory effect is different. The animals, while some are extremely intelligent, seem to be so much as a human, this will never be from an animal into the humansitic race. While it is true those who came down through the portal did take over the body and a portion of the animalistic soul, and the development and the change was then made from the animal into the human, this was a separate and totally different being and totally different structure that was then set up. This is separate. It will not, on the Earthly dimension, be the same as from one to the other. This is a beautiful thought though.

Many people feel that animals are making a lot of progression at this time.

That is true. Do understand that animals are making a lot of progression, and plants are also. Man has helped this much. Man has helped to develop animals and plants much. He has helped in many ways to develop these other soul entities. But they are other soul entities. They are different vibratory patterns and frequencies which are set aside from your own.

I'm glad we are giving something back for what we are receiving. In some books we were reading a few years ago, they told about skulls that had been found. There had been operations performed on the heads, in areas that we are not able to go into now. They know the operations were a success because the healing of the skull took place. Were some of the operations they were performing to still help the vibrations?

There were many types of experiments that were made, and there were also many types of operations that were given for disorders other than mutants. Many of those operations were given to people who had problems that were definable and many problems that were not. Some, such as mental illness, were not definable illnesses.

The "Others" that were there also helped since the brain was experimented with in the beginning and much was learned then. The "Others" learned much of your own particular animal brain and therefore, their knowledge was able to be carried forward to help the people then. They had made great strides and knew of some of the areas that could be worked on and could check those with their instrumentations and be guided in that way. Much was guided, as is now, with the tiny electrodes to be able to see into the areas, and to be able to stimulate the areas, and even on into advances that are not known at this time. For as known, the areas that were worked on cannot, even now, be tampered with. Knowledge is not great enough now to do work there.

New ways and techniques will be discovered to find problems and to check the brain. Experimentations for mutations were done but were mostly for corrections within that one body. As seen many times, there were different operations in different areas. This was not for looking for a certain portion but for different types of operations that were done.

Were the sub-humans used as slaves in Atlantis?

Often, and not necessarily by those of high stations like in La Mureah. No, they were not used as slaves in early La Mureah, but later on down through time, as time and customs and men changed, and their thoughts and ideals changed, man found that they could use them as slaves. Well, not all were treated as slaves either, in the term as you think of as slaves, but more or less as servants. As giving them

134

work and helping them. Not as a lower class of people, but as an intelligence. Giving them work and things to do so that they may continue their lives with dignity, and not have to grovel in the woods, or to fight to get their food. But they were taught how to live with the other people and how to do more menial tasks.

Those who were sent into other parts of the world, as they developed and grew and advanced, many of them used the sub-humans more as slaves and some were treated badly. This was an individualized thing though, and was not something that everyone did and was not accepted by most, and yes, some were used for experimentation also.

But the experiments were to help them weren't they?

Yes, mostly to help them. There was not experimentation to do worse things to them. This was for their help.

What about the slaves that were brought to America?

While this was considered a very bad thing and a type of thing that was unthinkable, these were strides being made. Do not look at everything as being negative. This may not have been done in the best way, but was done in a third-dimensional way for that time. It was done for the good of all, even though none understood what was happening, but was still done for purpose and has worked out well, even though many think not. Realize that these years are all years of special quick transition. Things are being done more quickly and perhaps things are not always done nicely, but are done in the way of the third-dimension, and are being done to make advancement quicker.

They have advanced more since that happened than they would have advanced probably in 500 years where they were.

That is correct, and this must be held in view when looking at this in an overall view. This is part of the progression of all. Much has been learned by this. Much living together, and understanding, and progression of each soul has come from this very thing. Times have changed and mixing and bringing together is now being done, but in another way, better for the times perhaps. Much learning comes from the mixing and living together of different peoples. Much energy and time is wasted on the embitterments and upon the right and wrong of nations and countries.

If the energy was put in another way we could probably advance three times faster.

That is correct, and in your United States is a development of great importance at this time. For there you will find the intermingling of many different types. Different peoples brought to one land to live together, and to be together, and to create together. This is of great importance, and this is bringing the vibrations up a great deal, and very fast also, very fast in comparison to what it has been.

In a sense, didn't war play a big part in this?

This is correct. Things are not always done in a nice way, also not always done in a nice way for a reason. Much karma is released so that people may make rapid advancement. When so much karma is released in a short period of time, for so many, there are many things that do not appear to be good or right. Do not misunderstand, I am not for any kind of hurt or suffering, but when I look at it as an overall view and an overall picture and can see the advancement that has been made, then I can say that this was a quickening of many things.

It has been known for some time that things needed to be quickened as time is running out for many, as has been told. Also realize that war does move a lot of people around that would not have been moved otherwise. Things must be done in your third-dimensional world as they can be done there, in keeping with your own ways to be done. Interference cannot be given. There is always something good that comes out, even of bad things. All is learning.

I would like to know where La Mureah was.

La Mureah was in the part that would be called the Pacific

Ocean, towards the Southern part. Many small islands are in that area now and is not of great importance, that was so long ago. This was basically where the portal was.

Is the Devil's Triangle connected with Atlanteah?

Yes, this will be given later.

Another thing that we have been wondering about. Before we came onto this Earth, had the large dinosaurs disappeared, or were they here when we came down?

Before and during.

They were here when we came down then?

No, not all. When the portal was first opened, where they may look through, this was of a very early period and lasted a great time. This was not just something that happened — two weeks later souls started going through and in two years it was over with. This was a very long period of time in through there, and as I said before, time was not then as time is now. This was all very different then. This was to watch and to see the forming of much of your Earth. This was not as looking from another planet or another place — this was from another dimension of Earth.

This was looking from another dimension of Earth?

Of course, your Earth has many dimensions. Many dimensions of vibratory effects. This was not as looking from another planet down onto your Earth. It was not looking down, or up, or around. It was looking through, looking through as if you can see into the Astral world now clairvoyantly. It is all within your Earth's atmosphere, within, around the Earth. This was from one dimension of the Earth to the other dimension of the Earth.

Then we were in this Universe already?

In your Earth's dimensions, Earth's dimensions itself. You

were not on another planet — you were of your Earth but of another dimension. While your Earth was being formed and while this was going on, there was more than just the plants and the animals that were beginning, that were taking this place as residence. There were other dimensions that were coming into being and the whole thing was forming. As told before, this all formed and came into being at that one time. It is as interwound with each other. You were not in the third-dimensional world. You were in another dimensional world which is around and intermingled with the planet Earth.

If we left this Earth in a spaceship, would we go through that dimension?

Of course.

It's just that we would not see it, or know it?

No, it is not a third-dimension. You only know of what is in the third-dimension. The "Others" who went there were from planets that were also in a third-dimensional frame and they are of the same type of animals, and these things that were on your third-dimension part. They had knowledges of third-dimensional workings. This is why they were sent there.

If I were of the proper vibrations could I see the other dimension?

You could not, no. This would not be a normal thing from one dimension of that type into another dimension of your type. This is why, when the portal was there, it was like a rip or a gap or something that was not normal. It was formed through a vibratory effect that was not a normal thing, which allowed a being of one dimension to see into the other dimension. It was not as a corridor that you would slide down, or anything of that type. It was as though these vibrations between the two had intermingled and had become more plastic, or as a soft material that could be penetrated from one to the other, such as cobwebs — although a barrier and something that is seen, it was different than if there was nothing there.

It would still be where you could pass through. Does that help at all?

I think so. But I would like to understand more about what you have given me. I know we talked about one ball inside another ball.

Like sub-dimensions and the other dimension is another ball around the first ball.

Then when we are looking out to the stars we are really looking out through the other dimension in a sense?

Yes, but you cannot see that, because your eyes cannot perceive it, nor can they perceive you. You work together but separately on totally different vibratory levels and effects. Totally different types of entities. Totally different types of soul developments now. You want to go back to where you came from? You were there once. The curious minds of the soul developments brought all here. Now, because they do not remember, they do not know. Now the curiosity is to go back and find that again.

What was the power source in Atlantis and La Mureah besides the Sun?

Much has been understood about the Sun in the past. Much was known at that time about their own powers that were generated. This was of early La Mureah and, of later times, the crystals and the Sun were found to be used in new ways. There were also some devices that were brought in by the "Others" that were much of how you would describe a battery type, but not exactly, but of an illumination that did not use either the Sun or any other electrical type of charge that you now use. They were used for lighting ways there when the Sun went down, as it was not dark, as it is there now. There were ways of lighting used at that time that was from the crystal.

Much was learned and much was used through crystals and not just from one crystal as is told of in Atlantis, but of many crystals placed in many different parts, in different areas, for different purposes. This was as you use now your motors and engines. It was used quite freely. Between the Sun and the crystals, there were also other inventions and other things that came up, from time to time, that have not been rediscovered yet and were not great discoveries at that time, because they were not needed so much at that time. But before too long, you are going to see, besides the lazer beam, which is being utilized now, but you will also see another invention come up quite quickly now that is going to revolutionize quite a bit of your thinking. Quite a bit, and it will be a light source invention.

A light source?

That is correct.

You say quickly?

That is all that will be given now. Just watch, you will see.

Will it be given freely?

It has already been given freely — how else would it be given?

Can you tell me more about the power sources in Atlanteah?

In Atlanteah the crystal was still much used there, but not anything like in La Mureah, because Atlanteah, as you have already said, was more primitive than La Mureah. But in Atlantis, after a period of time, as more attention was given to that one area, then this was more utilized, but was not developed in the same ways as La Mureah, but was more centralized, as has been stated about the Great Pyramid, the Great Crystal that was there, that was used. That was not just one great one — there were others also, but it was not in small use around everywhere as in La Mureah.

One of the reasons why I had asked you about the animals was to find out if the lazer beams were used to control the large animals, and did the lazer beams go out of control and destroy the cities?

140

Well, that is not correct, no, but there was much done to control the giant animals.

But were there lazers?

There were energy fields that were set up that were effective for some, and there was also much using of the natural forces. Much of the technology then was using natural gas that was found that would ignite. There was much of that used then. This did cause some disturbance when they were exploded, and the full knowledge of what was being done was not understood. Much more damage was done than had been expected. There was much done that way, yes, and also in Atlantis the Great Light was used there. The Great Stone was also used for that and was effective at some great distance, but was not concentrated on this one thing at any time, but was used some when absolutely necessary. This was not just to kill all those creatures but was for protection.

More to hold them out than to kill?

That is correct, and do not think that all that was done was to kill. This was mostly to stun or to aggravate or to turn away.

When in Atlantis, we grew food in areas that were under roofs, like the greenhouses we have today. In one of Linda's past lives, she was in the greenhouse and she had to leave quickly because the light from the pyramid was coming into the greenhouse.

That is correct.

Can you tell me the purpose of this?

The purpose was for nourishment and for light growth.

O.K., was there a reason why they grew under roofs at that time, rather than out in the open like we grow things now?

Not all was grown under roof, but much was for many reasons. Seasons were different then, things happened that, many times, were unexpected and many things in Atlantis were protected in some way or another. Also, when using the ray, the pyramid ray you call it, the Great Crystal light, the using of this ray also did affect and cause a chemical reaction when this process was used. It was not harmful, but was not good for the atmosphere. It only lasted for a matter of a few seconds. The element that it produced was very quickly dissipated, dissolved and was gone and did not linger. But this was also done to protect that from escaping into the atmosphere at the time — to contain it until it was neutralized.

Well, were they where light could normally get to them or was the only light they received from the crystal?

Oh, no, light could get to them, but remember, the seasons were not as they are now, and much was grown without the benefit of Sun, as always known and thought of as now. While the Sun was prevalent, it was not always accessible where things were grown to produce food. This does not mean that there was no Sun or anything of this type. It means that perhaps, during a small part of the day, as she has said, things were grown up the side of a hill or a mountain, and therefore it was not accessible to the Sun all day as was desired. This was even for crops that were grown in the full sunlight. This was an energizing, nutrient type of a thing that was utilized for healthy plants and better food and more abundant crops.

Are we getting close to knowing anything like that now?

When the development comes, you will know more about the Sun and the crystals, as was stated earlier. This may be rediscovered then.

Are there any of these knowledges in the pyramids?

Much agriculture knowledge is within the pyramids, yes. Much knowledge of different things, that was at that time, was put away for the future, yes.

Will I be part of bringing some of this through, in other lives?

That is correct, but not all.

Why were souls given so much free will on Earth and other places?

Not always are you given free will.

Can you elaborate a little more on that?

There are times, and there are places, where you do what is necessary to be done. While some free will is exercised almost everywhere, there is hardly a place that is totally without free will. There are many places where there is very little free will, and you must do what you have to do. This is for progress and for knowledge, for learning and for understanding. Most places do have different degrees of free will, and, as said before, your Earth has much free will, much free will. This has been utilized there as this is what has been developed.

In the way the Universe looks at things, has this been a good experience?

This has been an interesting experiment. Much fast progression can be made there, if utilized.

I've had a great interest, most of my life, in space travel. Is this for purpose as I progress?

Things like that are possible. Things like that are possible before you progress. Do not consider that as progression, or not progression. That is part of whatever is going on at that time. If you were within a world that had spacecraft now, that would be your experience now. If 5000 lives later on that same planet spacecraft were an old thing, and were obsolete, it would be an old thing there. This is just new to you now on the Earth, but does not particularly have to do with your soul's progression, only of your existence upon the Earth. There are these different experiences that you are anxious for now, and there are others who would be as anxious for an experience where there would be a wheel. This is all only a matter of the immediate progression of your desire and anxieties at this particular time — for your curiosity, for your advancement now.

I guess on this Earth we look at that as advancement because we are striving to go into outer space.

For the Earth's experience this would be advancement, but you are not always on the Earth. You have not always been on the Earth. You have been other places that have different types of progressions — on progressions where spacecraft would not, have not, and will never ever be thought of, or heard of, or known.

Have I ever been on a progression where there were spaceships?

This would be true, yes. Many that are there now have been places of this type. This is not all science fiction. This came from somewhere. This came out of memory.

I request that this connection be broken now.

Chapter Eight

(Tape 18)

Can you tell me a little bit more about why we decided not to allow the memories to accompany the baby when it was born?

In the beginning there were many things that had to be corrected and many things that had to be tried. When it was found that souls may enter into the bodies of babies, and when this was perfected more, it was found that knowledges were being kept and given to these tiny infant creatures. This was not desired. Can you imagine the knowledges and understandings of the adults of that time? Of the great knowledges being put into a tiny being who could not speak and could not control its bodily functions? It made many —perhaps mental problems would be a good enough word for you to understand — many problems of growth and many problems throughout the entire lifetime. For this one reason only, it was then decided that this was not desired, but there were other reasons also. It was found that, as this was carried forward, it created many feelings of guilt, or much confusion into the different lifetimes that followed. As the original energies decreased from one lifetime to another, it also made more problems.

Some memory is still carried forward, and often is seen in those who are young, who are able to do things that normally are not done, or to remember things that could be brought over, and often as you are finding now, as vibrations are being raised, is being done. But during that time period, as the vibrations were being changed and being lowered, it was found not to be a good thing to do, so it was readjusted and changed.

145

Total memory loss is not within a baby. A baby still has thoughts and ideas and understanding, and even many young children have more understanding. This is gradually lost, so that it is not brought into the adult life, as bringing through problems and remembrances of what has happened before.

The more lives you lead, the more things you have done wrong, which could lead to much depression. So, it is cleared off between lifetimes and you get a new, fresh start. These memories are not lost within the soul development, but only within the conscious mind so that you may proceed and go on, and progress as quickly, and as well as possible in the one life.

But, in the beginning, they did try allowing them to have the memory?

Yes, this was in the beginning. This was first done, of course.

Well, I want to get that point across, that it was tried, but it was found that these problems came up.

That is true; there are many other problems with that also, that would not be of real value to you. That was the great consideration for your understanding now. But there were other problems, frequency problems, vibratory problems, and other types of problems. But, this one of memory was the basic problem that they were unable to deal with properly, except with the changing of that one thing.

Inventing things is always in my mind, but I find I can't do everything in this one lifetime.

These things may be worked on by you between lives, and will manifest in another time. When the time is right, and all things are within the correct proportion, then what has been thought of previously and worked on previously by you will be brought into manifestation. This is the working of the Natural Law. Do not think that Thomas Edison only thought of electricity within his lifetime, or anyone else that worked on electricity. This was worked on for several lifetimes. Thought of, wondered about, strived for, tried to develop and even between lifetimes he worked on it.

In the beginning, did we start entering the body before we made genetic changes?

Of course not. You mean into the animalistic body that has been taken?

That's right.

That particular body was not entered into easily. This was with much difficulty. Some of the other bodies that were offered were much easier to get into, and out of, to do experiments with. This particular body was a very strong personality and its own forcefield and vibratory effects were not in the best compatibility with the soul developments that came in. Much work had to be done and many changes had to be made. Much development was made within the animal body before the actual taking of this body was complete and satisfactory.

So we didn't enter the body actually until after we had made some changes?

Not into this particular one, no.

What ones did we enter into?

This is of no importance. There were many that are not even there now, you would not even know what I say.

You said, at one time, that the first ones that came down through the portal were more the scientists. Did you mean scientists as we understand them today?

Not necessarily, no. When I speak of them as the scientists, these are words that you could understand. They were the ones who looked at the situation and said "There will need to be changes." Some wanted to play and do other things, or prepare the place that the soul development, at that time, wanted to live. There were many things that were done. But some of those who came through said, "Let's try to figure this out, let's try to get into the swing of things here, so we can learn more," that is all.

This was not a group that were called scientists that came down to try to do this with that purpose in mind, no. It was as though a group of people would go somewhere together for an outing, and some would gather wood and some would make a fire and some would swim and someone else may take care of the children. This would not be as a scientific community as it would be known today. The ones that began to do the experimentations and began trying to do things were a more adventurous group, but were not as scientists from one dimension that went to another dimension, no. That was not the case.

When we worked on the genetics and things of the bodies, was this done in the womb or after?

Much was done on the adult population, much was done and worked through thought forms and through changes of electrical and the principals of the aura. Of course, in the womb is an excellent time to make genetic type changes, and to try to aid in the development of the fetus. This, of course, would be the best thing to do.

There were also changes made with the adult, because as changes were made with the adult, then changes could be monitored in the babies that they produced and also much could be speeded up in each adult lifetime. It was working with all things that were available, as much or more work had to be done with the aura, to make changes with the physical, than even had to be done with the physical.

Then the aura has a temendous amount to do with the physical?

Of course, even now. The animal aura is different than what is now the human aura. The animal aura was more in tune with not only the animalistic soul development that was going on at that time, but also with the things that were within the Earth's atmosphere that came within that animal development. As the frequencies were changed to handle this new soul development that was coming in then, the aura had to be changed greatly. It is not the same as an animal at all. Many more technical things are added to the aura to make what is a human. Many more intricate parts were utilized. It is a much finer and different type of mechanism.

We see auras of other people, is it possible to see auras of the animals?

Auras of the animals are there. They are not seen a lot, just as the auras of people are not seen a lot. The frequencies are different. Your eye frequencies are set up for certain things. The aura can be seen through the eye, but can also be seen from within the soul development, or within your aura that is transmitted back. It is not only with your physical eye that you see, and I cannot say that it is from your third eye, as is said by many. While that is a way that can be understood by many, that is not correct. This is something that is being developed and being brought up by the vibrations, as many more people are being able to begin to see auras, and as refinements are made you will be able to see more.

Then as we go along we will be able to see more of the animal?

If so desired. The etheric bodies are on a much lower vibratory field level, as are the bodies of humans and animals in your third-dimensional world. The etheric double, as it is called, is seen by almost all, that is because of the low vibratory field that it is in, which may be seen by the eyes. The eyes are designed to see your third-dimensional world, is that understood? So that is why the etheric body is seen by so many. It is when you get into the finer vibrations, that the eyes are not made to see, that it becomes more difficult.

And that's the reason that, when you change your vibratory level, you can see more of the aura? I know that Linda says she sees much more through hypnosis. Does that mean that she is changing her vibratory patterns?

That is correct. You are changing the mind from which you are observing.

Is it possible to duplicate this artificially?

Not 100%, no.

They are now photographing auras in very small areas, such as fingers.

This, that is being photographed, is not of the aura. This would be of the electrical fields and the electrical currents. Perhaps you would call it the aura, but it is not of the true aura.

It is the energy around the body?

That is correct. More the etheric energy or energies of the body. Not of the working aura.

In order to photograph an energy around the body, your film would have to be of a certain speed wouldn't it? In order to pick up the vibratory effects?

The film would need to be of certain speeds, of certain light variations, and also certain processes on the film. There are many technical devices that can be made to do this. At this point things are being created. This process that you speak of now, is being utilized and will be more perfected. There are constantly more advancements being made, small advancements perhaps, and taking time, but still advancements.

"This One's" guide, Marcheah, I understand, only took a body once, which means she retained her knowledges and vibrations. The ones that are not of the same vibrations as she, their vibrations will not allow them to go on that level, will they?

That is correct.

Then as we progress, even after we leave our bodies at death, we are only going to the levels where our vibrations will allow us to go?

This is correct. This is known and this is correct. This is a truth.

If I wanted to travel to your dimension are there certain avenues that certain vibrations allow us to travel? Or am I getting too complicated?

You are getting complicated for now, but it is not as an avenue or as a street.

But it has to do with our vibrations?

It will have to do with your vibrations, yes. Your vibrations allowing you to enter into certain areas. There is not just one place you can go, but there will be many choices you may make, and you may make those through your vibratory patterns.

There seems to be a lot of channelling going on now. More than in the past. Is there a reason for this?

Many are working at this time to raise their vibrations. Many on your side, and many on the other side, are working hard to do what they can to raise their vibrations. This is very good, but as you have already been observing, the quality of answers that come through are not always what is desired.

"This One" knows that she is not allowed to do channelling, and it has not been done in her other lifetimes.

That is correct, she has not. "This One" knows what is best for her and, because of this, she is now still able to contact this dimension.

When she came to me in Atlanteah, and I helped set this connection up, so we could work through her, I was not sure what was totally done on your side.

Understand this. Much preparation has been done to keep this open for her, if possible. Much has been done by others. Marcheah was chosen, a very long period of your time ago, to

151

be her guide, because of the frequencies and the nature of her understanding of all. She is trained so that she may communicate also with "This One", but not from or through the body. "This One" keeps her connection with all at another distant point. Marcheah may connect with her similarly, as I do, but on the Earth dimensional level that they would work from. As you know, the teachers I have given to the two of you, that "This One" can communicate with from my dimension, are all gone basically through me, and met at a place where she may have entrance into these dimensions to communicate with the proper person. As others may go through the same dimension or they may go through the same area and never communicate, as they do not know where the door is, or how to knock, or how to get through into that dimensional level. I have given the symbols that make the passage for her available. If you ever wish to do channelling through "This One" you may do so through Marcheah. All information that I give to you is given freely and may be given freely to others, with reservations. You may ask and I may give permission, but basically this information that is given to you is to be given freely, especially to those who have understanding.

As you give me knowledge and my frequencies are changing, when they get to the proper level, and "This One" brings me to you, then that connection will stay, no matter what I do? Or do I have limitations?

That is true, this will be explained later.

What do I tell people when they want me to prove where my information comes from?

I do not need to prove who I am to anyone. I do not need to create something to happen to prove who I am. Many do need this, many do desire to be proven. This is not necessary for me. Those with higher understanding, and those who are looking for the truth, are the ones who will have the understanding and will have the knowledge and these are the ones I am trying to reach. I do not need to change those who believe in Jesus. I do not need to change those who are beginning to have understanding of Metaphysics, but do not have the ability to go into deeper parts of Metaphysics. Those who are on a level of understanding that do not have the capabilities, and many do not have the vibratory effects that give them the capabilities to understand the higher teachings, that is fine also. Teachings are for higher knowledges and understandings for those who are truly looking and are able to absorb and understand the higher truths. I did not prove myself to you and I do not have to prove myself to them. It is a waste of my time and energy and of "This One's" time and energy. There are passages in every great religious work that there is upon your planet today that tell not to waste your time on those who do not understand or cannot understand. This has been known and is not just said by me, but this is a Law of the Universe.

We would like to know more about our guides.

The guide that is with you, is for you.

Yes, but a lot of people contact other guides.

All guides that you contact are for you. In other words, they are all of the same type of vibratory frequency that you would be able to contact. It should also be understood by you that you do not know what their soul development frequency is, by someone's guide, because a soul development, such as you, could be of very high frequency and could contact anyone from your high frequency down. Or this time could be given a guide of a much higher frequency to work through, to bring that soul development up. This again is more individualized. Anyone that could come through into your frequencies that you allow could be your guide if you ask them. They may not be your personal guide, but they could come and speak through you, if you allowed. This would be permitted.

But you wouldn't want to leave yourself open to just any guide, so the only way I would do anything, would be to trust the feelings of my guide, to lead me to another one.

That would be a good way to do it, if you didn't wish to use your own personal guide. Someone with a lower vibratory frequency may not be able to contact you. They may not be able to come through to your development. So, it normally should be someone that is close, that can come through your vibratory frequency.

That you can handle well, and they can handle you well?

That is correct. Or else you would have to go down to their frequency. You can go down to their frequency, they may not come up to yours. You see, this is as most things that are within the Universe, the lower soul developments may not intrude upon a high soul development. They may come down to work with you, but you may not intrude on them. It is not because they say, "You are not as good as I am, and you cannot come here," that is not true. But, it is as though you cannot get through to them, as though there was a barrier, until your vibrations are raised up to match theirs.

Is what our guides learn limited to what we learn?

No, not through your learning. They make progress through their learning. Whether you learn or not does not have anything to do with their learning. It may progress their learning, but if you do not learn it, it does not hinder them.

But can they actually raise their vibrations by teaching you?

Of course.

It's just that, in that case, by being a guide, learning is slow isn't it?

By being a guide? Much progress is made by being a guide, but you cannot make as much progress as fast normally, by being a guide, as you can with a dimensional body. Because you have the opportunities to act out and to interplay and create your own situations.

But why do different ones decide to be guides, rather than come back?

Because, for one reason, they all cannot come back at once, there are too many. Second, there has been set up that there must be a guide, at least one guide, for each human that is there. Therefore, you do have the guidance from the other place, at least from one direction. But there are many, many more there, than there are on Earth. You sometimes take the one life and guide someone for the experience, or for the knowledge, or that may be a special friend of yours, there are still feelings. This does not end all feelings when you pass on, and sometimes this is desired.

To help one another?

That is correct, to help one another. But this is not always of great importance. Mostly when you leave, you leave. When you leave, there is no more importance there. You go on about your other duties, and your life on Earth is as a dream, and this is then forgotten, or this is put out of your mind for that period of time of adjustment or relearning. Then the time comes that you have the chance to go back into the body, or to do something else, or this type of a thing comes back into play, or what level of help would be best for you at the time, as seen from that time, and seen from that dimension. Just because someone dies, and was an Indian, does not mean that they are going to be a guide now.

It almost sounds like dying is the beginning.

This is. These are two separate, different worlds, that are intermingled with each other. Two totally separate existences that give you the creative ability that is not always found in other places. Other places are more — not all other places, do not misunderstand — are more of a one type of dimensional learning. The Earth has many avenues of learning and understanding and creativity.

How are guides selected?

All are given the opportunity of being a guide that is of a certain level, of degree of understanding. That means that they are not of a development that would — still would be wanting to — I'm not sure how to phrase this so I'm not misunderstood. Perhaps "play tricks" would be a good word. Of a soul development that is an honest, sincere soul development. Soul developments are of many different levels. Some soul developments are still in a series of forms that would be thought of more on a baby level, childish level, intellectual teenage type, and again those are only as a description, an idea so you may understand the level of development they are still on.

Some souls that have come into your dimension have not taken that many different lives to learn. They are comfortable, they are not pushed, they are given opportunities, as are all. They do not wish to take the step. Many of these are even higher soul developments that are not that adventurous. Many of these will still be able to come into the Earth's atmosphere as the changes are going on, and afterwards, because they are of high development. But they are not developing themselves here and now, and so are more on the baby type level, as you would consider someone who comes back and learns lessons and works their problems out. They are immature on the progression level of the third-dimensional thinking. Perhaps that could be understood better.

So friendship does have something to do with who your guide might be?

Often, yes.

Such as Linda and Marcheah?

Marcheah was chosen from the beginning, as this was going to be her choice to continue from her dimension. Basically, because she still had the original knowledge and did not lose it and is helping much. She is very highly progressed for help there, as would be needed in any job for top level supervision. When those who came down and were given the special assignments, such as "This One", many special knowledges and special handlings were given to those that were chosen to be their guides. The guides were then chosen, at that time, so that they may work with these that came down.

The connection has been between them all through the time until the connection was lost and then was no longer needed. Then the guide was free to go on to do other things that it wanted to do. So, the connection between "This One" and her guide has not been broken, and Marcheah is still with "This One". So, this is the relationship between "This One" and Marcheah. This is not from friendship or from attraction of Earth lives.

Kind of like a job?

This was like a job in the beginning, as both were doing the jobs they were prepared to do. But, of course, as with any job, the longer the attachment is there, the more the friendship is. This is not different.

That's true. Well, I notice in some of the past lives that I've done with other people, they have been guides of their guides.

That is correct, that often does happen. And many times guides do stay until they fulfill themselves or until the people fulfill themselves above the guides and can no longer experience together. This often happens, on both sides.

I feel that I've spent most of my time coming down to Earth.

This has been of great importance to you.

Sometimes, after the death of famous people, there are people who claim they can get in touch with them. Is this possible?

It is more possible that you would be contacting the astral shell or the energies that still are from one of the bodies, but not from the actual soul development. This would be individualized but probably not connected with that individual soul development.

Could you actually get information from that shell?

As long as the energy is there.

157

Some people say they are in touch with certain doctors who lived many years ago, and these doctors are helping them do operations.

Sometimes, when the time comes that they are ready to make the choice, they do come back and act as guides, and they give their information. This is much exploited there and information given is not always of the best quality.

That's what we were wondering, how much truth there is in it.

Only some.

What about guides on different vibratory levels — can they communicate with each other? Not go there, but just communicate with each other?

This is about the same. In other words, the frequencies and the electro-dynamics of this other world is different from yours. Where you are, you will use thought and you will go to someone's house and knock upon the door and they tell you to come in, or you call them on the phone and they acknowledge you and give you permission to talk with them. Now, on the other world this is not done in the same manner because the third-dimensional personage is not there.

There, things are more on a vibratory effect and, if someone wishes to communicate with you, the vibration is set into motion and if you do not wish to accept this communication you do not have to accept it. You have a privacy. While all are there together and interwound, you still have your privacy. It would be like being in a room of thousands of people, but still be an individualized private soul with not hearing, or not knowing what is going on, unless there is a communication between you and one other soul or ten other souls, or however many wish to communicate together. But, only if the desire there, is from all to communicate. This is a different thing that is going on there. So it is separate from how you communicate, verbally. There is no verbal communication there.

Is it all done with energy?

Yes, this is all done with energy. These are the dimensions, think more about this now — about your dimension — the dimension where there was the opening, of what I called a portal. Now think of that coming into the third-dimensional world, and you have more of a basis to think upon. So, that may help you two ways.

I've always thought that one of the things we take with us is our thoughts, what we create here. Can we draw on our memories there?

The thoughts are taken with you and the thoughts may be used to create the place that you wish to abide in. The heaven or the hell, where you are, what you are doing. This is part of the thought-form creation, and since you have been in a third-dimensional world you may create a third-dimensional world around you. This is only for you, unless you wish to bring others into it.

Is it possible to bring others into it?

It is possible to bring other people into it, or other souls into it. Now this does not mean that the other soul is having to take a form and coming into your thought patterns, into your area. There may be many dozens of souls in the same area creating their own thought patterns. This is strange for you to think of, I understand, but this is still within that dimension. That is still possible there. There is no dimensional effect as in your dimension. So, matter there may be utilized in more than one space.

Much as we can create going back in our past life?

As though you create it, as though it was real for you then.

In other words, feelings and everything? Because when we go back into a past life, things are as they were at that time.

That is correct, only you may create these things in thought, as though they were real upon your own dimensional world now.

Is it necessary to have that experience on Earth in order to think of it and create it there?

Not necessarily, no. You may create there, known or unknown, you may create whatever you desire.

When you say there, do you mean directly after we leave Earth?

After you leave the place that you are now. Where you are now, only one piece of matter may be in a space at one time, two pieces of matter may not. But there, many pieces of thought-form may be in one place at once and yet none intrude upon the other.

That's very interesting. I have a lot to learn.

That is true. You create the thing and the people that you would desire without that soul actually coming to participate. Although, it would be very real to you, for as long as this is what you desired.

Is this the reason why some people who have actually died for a matter of minutes, and then returned, say they were with their loved ones or a spiritual being?

No, this is not the same. As it is desired, it is given to these, so to speak, by those that are helping. In other words, the ones that are newly deceased are given what they expect, and what will help them, by those who are helping them with the transition. There are those who choose to be guides to help these souls come across, because as there is birth and death on Earth, there is as birth and death in that other dimension. Not the same as on Earth, but still a new merging into the new place and a going from that place, and so there is a transition period. Just as there is a transition period into a baby until that is accepted and is utilized. As dying is usually a slower process, unless by accident or something that makes it immediate, so there are those who help you go back and forth. This is what they want to do, and they often can create from the mental pictures and images, to help that soul understand and to accept the new world and to be comfortable. Do you see? What would happen if, all of a sudden, you are somewhere strange and you see strangers coming toward you? Would that make you comfortable?

No.

No, of course not. But, if this was someone that you knew, and loved in the lifetime you just left, your mother, your brother or someone that was of emotional tie to you, this would be accepted and make you happy and full of joy.

I think that's nice.

Of course, and this is a way of protecting and helping.

In other words, we actually try to help one another?

Of course that is true.

> NOTE TO THE READER: In the time spent with the Guardian For Our Universe, we feel there is tremendous love and understanding on the other side.

I'd like to know a little more about the soul's progression on Earth. Can we tell who is making progression?

Those that are on your level, that you think of as digressing, or not making certain progress at all, this is, as said, not necessarily true. Although, it may appear from your observation to be true.

It's just a role they are playing?

Sometimes, and sometimes they are not making much progression. Sometimes they truly are not, but this is only from your observation, and sometimes there are many who — probably more — that appear to be making progress, that may not be making progress. In other words, they may be at a place where they are at that time, and appear to be very high soul developments, from your observation. They may, in all truth, be very high soul developments, but they are not progressing themselves up above where they are now. Where someone appears to be a low soul development and is working at it, they may be making progression, because they are working at it. The others may just be coasting. This is why it is very difficult for you to make other than an observation on your part.

Would it be possible for the President of the United States to be of low vibrations?

I would say, that if you are using the President of the United States, of your Country, I doubt that these people would be of a low soul development. Ones that have obtained this position within your Country are people that are developed. Not only financially or in Earthly ways, but are souls that are trying to develop themselves, and have developed to a point where they can handle the extreme energies put upon them in this position.

This is not as in other countries where there are only those born into being a King, or a Queen, or into a high station. This is truly why the soul that wishes to be born into that position should only be of a high development soul, because they know the position they will be born into, but this is not always the case. The body may not be suitable for a higher soul to take, and to work his best development for governing people through. It may be a good body for a soul to take to learn on the soul level, but not necessarily to govern people, so the preferred leader soul may not come through.

This is why the elected portion is preferred, because you may choose someone who has a chance to be of a higher nature. Someone that is more suitable to lead you, and not just take whatever happens to come along.

Thank you. You may release "This One" now.

Chapter Nine

(Tape 19)

Will you please explain about going to God, so I can understand it better?

You have been taught about God as a duality, as two. As you are one and God is one, and God is supreme. Is this the God? There is no such God. You know this has been stated before. There is no entity, individual person as a God that is a duality with you. This is a misconception for many reasons. The more specific thinking of a God consciousness, as a consciousness of a thought, of a fluid type of energy that is within your Earth's atmosphere or Universal atmosphere, this is a more correct type of thinking. Although, it is not totally true, but much closer to the truth than the other. The other is not even within consideration. The other God, that is thought of as Universal thought and lifegiving substance throughout the whole Galaxy, Universe, is more true. This is the God that is of "The portion that is called God." Is that any help?

Within the ethers of your world, there is a substance that nourishes. The air is a substance, as the water is a substance, this is understood there. Much comes through the air to nourish you in your body. Much also comes through there to nourish and to help the aura, to be intune with, to bring energies through. This is part of the workings of the aura, to be in contact with the portion that is called God. That portion that is called God is manifest in every drop of water, in every atom, in every cell of your being. It is everything, and it is not one thing, so it could be called nothing. It is everything, but it is not of one individualized form of itself, otherwise it could not be air and water at the same time, it could not be in a person and in the tiniest and in the largest. It absorbs all and consumes all, that portion that is thought of as God. That is that portion. Does that help any?

Yes.

It is a creative intelligence. It is a vibratory effect. It is a nourishment. It is extreme intelligence for the beings of your Earth dimension and of your galaxy. More knowledges than you have permeate basically from this portion that is called God. And this is why it is said, that it is more of a feeling, it is a feeling of how you were in La Mureah and Atlantis. It is a feeling of the oneness, the more completeness, the more understanding, because at that time you had thoughts and powers and understandings of the true "you" brought from another dimension which is more than you have now. So, to get back to the feeling, is wanting to get back to God.

Getting back to something, but it is not a thing, it is not a person, it is not an animal. It is not something that is compatible with you as an entity being, or as an animal or a person, it is not that at all. It is this all-permeating thing that creates through vibration, and has understandings and super-intelligence for you. This is the portion that is thought of as God. An all-giving, all-loving, all-vibratory experience. To become this same type of loving, understanding, vibratory effect is what is desired at this time. To become one with this vibratory effect.

You may be able to understand now, if you understood the other day when we talked about guides. As a guide that you can go up to a certain level, because that is where your vibratory effects stop. You may communicate with all those below if you choose. Your vibratory effects can go down to theirs. You understood what we discussed about the guides on the different vibratory effects.

Yes, I did.

This is it. This is not because God is perfect, and you are

not perfect. You are perfect where you are. You are perfect upon the level that you are perfect upon. When you get upon another level, you will be perfect upon that level. When all things match and all things harmonize, when you learn to harmonize your vibrations with this portion that is called God, to bring your vibrations into correct proportions, this is your learning experience.

Your learning experience on your Earth is not to be good, or to be bad. It is not to see how nice you can be or how ugly you can be. The fact is, being nice will help your vibrations. It will make them in the more correct mode of what you want. When you learn on your dimension of Earth, that you are on now, how to put all your vibrations into an effect where they have harmony and blend beautifully together, and make the musical concert of your vibrations, that you may blend into the vibratory concert that the portion that is called God plays, then this is the lesson you learn there! Is that understood better?

Yes.

Color, harmony and vibration all play their part. It is all one. This is what I told you before, one planet you go to you learn one thing. If you have two things to learn then you must learn how to bring those two things into play harmoniously. On your planet you have many things to learn, to blend and to make harmonious. Many different vibratory effects that you have to learn to blend and put together and allow your aura to become the vibration that makes you harmonious and beautiful and acceptable into the portion that is called God. **This is your lesson there.** The little things that you do are the cause that makes the effect. The things that you do, as karma — let me explain this to you now, about karma, that has not been understood before.

When karma was explained to the third-dimensional world before, it was explained the best that could be told and explained for all to understand, which is fine. When you do something of negative force, a bad thing, then you get a bad thing back to you. It is not for everytime that you tell someone you hate them, someone will tell you that they hate you! No, that is not correct. It is a vibratory frequency of the anger, and telling someone that you hate them. The likeness of that vibratory effect must be accepted back to nullify that, so it makes a balance. When you put this negative vibratory effect into play, you must have something to balance that back to you to make it nullified. So, it is not when you kill someone, someone must kill you, no, this is misunderstood badly there. When you put the vibratory effect, the feeling, into that act, then something of equal vibratory effect is put back into your feeling, into your aura.

Remember when I spoke to you, the other day, of the vibratory effect around your body? Of what it will accept or what it will keep out? This is much the same, it lets in and it puts out or it will not let in. Therefore, you have what is called karma, but it is not an act, it is a vibration, as everything is a vibration. So there perhaps you can see that karma is vibration into your aura and into your soul being. Your aura is much protective of your soul being. When your soul being is in total vibratory effect with the portion that is called God, **then this lesson is learned.** This is a lesson of vibration there, this is not a lesson of what is considered good or bad in your dimensional thinking.

Say "This One" and I have trouble in this lifetime. Does that mean that I have to come back and live out that trouble with her somewhere down the line?

It manifests in the third-dimensional world as trouble. This is a manifestation of the vibrations between the two of you that must be null and void, that must be accepted. Good is done and good is given the same as bad. So, it makes no difference when you are talking of good and bad, this is not the point. The point is of vibrations, but do understand that what is called good vibrations, the positive vibrations, is what makes the beautiful melody that may be played within harmony. It is not the discord that is desired, but the beautiful harmony that is played within. When you may harmonize your soul with the portion that is called God, then you have learned a lesson, and how you learn to do it does not matter. It does not matter if you learn it on the third-dimensional world, or the world being a guide, or if you use the two, as long as you learn to control your vibratory fields, to bring your thoughts, your feelings, your emotions, all into this correct vibratory pattern of harmony and balance. You may learn it in one lifetime. You may learn it in a million lifetimes. Then you may pass through this portion that is called God, because your vibratory effects will allow you to pass on into something else. Is that understood?

I think I am going to have to go back to the tape and listen to this again, so I can understand it fully.

As I give you more on vibrations and bodies, this will become more clear to you and you may study the voice tracings also. As you come up with more questions, do as you do now, and ask. But vibration of harmony is the key to all. To go back to the portion that is called God is what you strive to do. As said, it is of good striving to do, but on Earth it is done in the wrong direction. That is why meditation plays such a very important part in what you do. Meditating does not erase karma. It cannot erase a physical act. It changes the vibratory patterns, you see? It changes the vibratory patterns that you are in, so that it will cancel out much negative vibration that you have. It helps you to be able to, what we call, lower the vibration, or raise the vibration of each thing. This is why I tell you it is so difficult talking of lowering or raising each different kind of vibration. Each individual person is trying to adjust their own vibration into that perfect harmony. If some vibrations are too high, they must be lowered; if some are too low, they must be raised, so this is why it gets confusing. This is why the meditation works and helps so much, to be able to balance and to be able to put these in more perfect harmony. Are these things now beginning to come into something that is more understandable for you?

Yes.

Excellent, but you see if we do not discuss, as best we possibly can, each individual thing while we are talking, then it is hard later down the line to bring this all into place, so that you may understand it as a whole. You see, as you are working with stress, vibrations and blood pressure, learning to bring this into balance, and into your feelings, your emotions, and your understandings, this is all an example of how it is worked, and it is all learning.

Yes, I know it is. And I'm also learning that what you think has a lot to do with what you do, and how you react to things.

This is more than you understand, upon your Earth. It cannot be overemphasized, it cannot be. This is one area that is not being utilized and it is there for your development.

Yes, I'm beginning to understand that more everyday. Some of the books that we have read say that, when you are having an out of body experience, there is a cord from the astral body that is attached to your body. Is this correct?

This cord is of astral vibrations. It is not a third-dimensional cord. This is made of the same material that the astral body is made of. It is of the astral projection and it is a connection that is into the body.

It is the energy connection?

It is the energy connection, and it is also the way back, so to speak. The connection that would help you find your way back, it keeps connected to the physical body. This is always attached until death when it is permanently severed and on the astral dimension is very important. It keeps you connected between the two worlds. It is your "portal," so to speak, between the two worlds. I will give you more on this later.

Can we direct our astral body and take our consciousness with it, to go places, to see things and then come back?

This can be done, but it is not easy. This does have some to do with the development of the soul, but this does not mean that only certain types of people may do this. It is just not as easy to consciously remember to bring this back. But it is done and can be done. Do not attach superiority or greatness to this.

In other words anyone may do this? Is this a matter of consciousness?

Yes, but not necessarily a matter of progress. It does not mean that when you do this, you are progressed to a certain stage, within your soul development. Things that were known at one time were not understood. The vibrations were of such that only some people understood them, or many were told of them and did not believe them. But the message always was given, and then as times progress, and intelligence progress, and vibrations increase, all these things go together, and all things are becoming manifest more and more. Occult things, they are called I believe, are being made manifest there and will increase as has been told before.

Are there any limitations to the distance that the cord will stretch?

This does not stretch, this is just created.

O.K., can you create your energy cord as long as you want it to be?

Of course.

Does your astral body travel as fast as thought?

Let me just say this. You are asking for exactness again. This exactness is only on your third-dimensional world. Realize, that when you are not within your third-dimensional body, everything is changed. Time and space are completely different. Even though you are still within the third-dimension you are in a sub-dimension of the third-dimension.

Things would look different to you?

Things are completely different. So, what you call traveling a long way in an instant on your third-dimensional level, may not be an instant on the other dimension. You ask me to give you an answer, I ask you, "In which dimension do you wish this answer to be given from?" do you see? From your third-dimensional level you may proceed slowly. In the sub-dimension you may be there within an instant, within a few moments of your thought.

In a book I read, there was an experiment on astral travel where two people were in the U.S. at a phone, and two people were in England at a phone. One person in England wrote something on a piece of paper, and the one in America answered it in a matter of moments and was correct. Supposedly, he saw what was written down by astral traveling.

But, do understand that in the sub-dimension, the whole vibratory effects change, and the whole dimensional thoughts, and beings are changed. This is why dreams are not the same as on your dimension. You may have a long dream that takes only a few minutes in the third-dimensional world, but the long dream was within the sub-dimension.

Oh, I see.

The time is not the same.

Is there anything that is dangerous to us, when we astral travel?

It is part of you, it is part of your being. It is as much a part of your being as your third-dimensional body is.

Then there is no danger there?

That is correct.

The other day, when we were talking about Jesus, I was going to ask about his teachings.

One of his greatest teachings was of love. At that time on your third-dimensional world with the understandings of the people, much was put into the third-dimensional thinking. It was told, and he tried to demonstrate it so people may understand with their third-dimensional thinking. Therefore, this was the only teaching that was given to the people to understand. But love is what changes the vibrations. That changes the vibratory effects that take you to God. These are the things he was trying to get over, to help people to understand the metaphysical things in a third-dimensional world. His task there was great, it was great for the people and the understandings they had at that time. He had much to do.

You see there was not much progression being made at that time and thinking was more on one level. Many were paganistic and there was much brutality then. The Jewish people were probably some of the most Godly type, at least in their thinking, but they were not progressing, because they were continuing with the old practices. They were just continuing on, and never making enough progress.

So, it was brought about that he would try to raise the people's thinking, which was not as successful as had been desired here. Those of whom he chose were good souls, but were just still of much dimensional thinking, and did not understand all of his teachings themselves. They were on too much of the one level and could not raise their thinking up high enough. So, it came about when the teachings were being written down, that this same dimensional level thinking was used. Even those who understood above the dimensional level, and understood the higher teachings of Jesus, still had the same problem he had, you see. They still could not convey to the people the meanings either, and they tried in their own way.

There were other things tried, things were told different ways to try to make the people understand. Much like I explain something to you and I feel that you do not understand, so I try to explain it another way. But so much dimensional thinking was put into it, that the words were spoken as "This is fact." You and I discuss often, "This is not fact, this is for your understanding." This made the Bible controversial. It is with so much Earthly thinking, so the people could try to understand, and to bring their thoughts and ideas up. Which eventually, and in a round about way, has helped greatly, but still nothing compared to what was desired.

This is why I have told you that people must change and not to continuously only look to one avenue of thought. The Jewish people still continue to look to the one avenue of thought that was back so long ago, that people could not understand then, above the very low dimensional level. This is not for now, they must change things and move forward.

This was the connection between Jesus and the Jews. It was not that the Jews were the chosen people or not the chosen people, it was not any of these things people think. They were the people, at that time, that did have the most thoughts, that Jesus could work through. He tried to work from there because they had the thoughts and ideas that could be accelerated. The people chose not to accelerate their ideas, but to keep them on the low mundane level as they had done for many thousands of years.

While, at one time, that was of good thinking to do, because it was needed for those people to stay together and keep their thoughts and ideas so that they would live on. Now they need to progress upward and expand their consciousness. Much advancement and much help was given through what Jesus did, as magnifying it and making it of importance did help the progression of the world much.

Oh, I can understand why.

But, it is time to get out of that time, to open up and to go on to other dimensional levels. You cannot stay there forever. You do not want to ride in an airplane that was one of the first airplanes made, forever. You have progressed on, you accepted that as a good start and you continued to expand and to learn more. There are many more advancements that are coming that will be quite shocking to you, but will be of great advancement. You do not want to stay on the same level, try to look beyond, and go beyond and expand your thinking. You must expand your thinking, so that you may understand more.

Is the healing of the body done through the aura?

The healing of the body has many processes, there is not just one. Part of the healing comes from the individualized cells that are within the body, from the energies that are within the body, and much of the healing also comes from within the aura. Much of mind is also in control at this point, and through the aura comes the energies from the external that are brought in and made manifest into the aura, and then into the body. This is as a complete cycle, and not as one individualized thing.

Can you, at this time, tell me more about how to direct the mind to the energies in the aura for healing?

Through thought.

Does it have to be through meditation or can it be through constant thought.

Define constant thought.

Having good thoughts would be a simple way to say it. Thinking of good health for yourself.

This is of good thinking, but also to direct this energy to the place, or to the area that needs help. To direct and to guide with your mind what is to be done there, to your best ability, will also be much help.

Did we all originate from one source, in the beginning of us as souls?

At this point, the understanding on your Earth is not so this may be told or described to be understood. There is much here that you wish. Too much technical knowledge that would not be understood by you at this time. Do know that all is continued and all is regenerated. I have spoken that as I get into the law of cycles, and into the law of dimensional cycles, these things will be explained. But, knowledge at this point is not sufficient for your understanding. All is generated and regenerated, and created and recreated, and is on a continual aspect.

Something that happened that everyone has been curious about, was about the Maya Indians. Their total population disappeared. Can you tell us more about that?

These people were taken.

Were they taken somewhere else?

They were.

Can you, at this time, tell us where?

They requested it. It is that simple.

Did they leave this third-dimension?

They did not leave a third-dimension, no. They left your third-dimension.

They went to another planet?

That is correct, "Others" did assist in this.

Did they leave here because they thought that another planet would be more suitable for them?

That is not correct.

Can you elaborate on why they left?

This was as "Others" wished this to be. This was discussed with the population and it was agreed that they would be taken to another place.

They must have been a very high vibration.

They were, and they also had much knowledge that you do not understand yet. Those people, at that time, were not as ignorant as thought. This is speaking as soul developments.

Then at this time they are living in a totally different —

Their knowledges were needed elsewhere and were taken by their free will to give service for others, much as "Others" came here to give service to you.

One more thing I wanted to ask you about was God. When we came through the portal, did the feeling of God come with us?

God is your word that you use to describe this. In the beginning before your Earth was there, and before many other planets were there in your basic solar system, galaxy area, there was a vibratory teacher that was there for these different places in that particular area of your universe. This vibratory teacher has been there for a very long period. The area that the Earth was then formed in, has many dimensions that were being formed and also many new aspects were being formed within it.

At that time, the soul developments that were of another dimension, that were able to look through that dimension, and see the third-dimension, were aware of this vibratory element. Although, not totally understanding the way to be in communication at that time. But were aware that there was a vibratory frequency. When coming into the third-dimension, then were their vibrations, understandings and knowledges changed. These understandings and knowledges came not only from the soul developments of the animals and plants, but also from what you now have termed as nature spirits, because at that time all were in connection.

As the development of the one body that was taken to house what would be called man, all of these new understandings were taken into consideration. As the new bodies were made, and the new vibratory frequencies were formed, it was also taken into consideration how to connect and to get into this teacher's vibratory effects that were there. This was taken into great consideration when the aura was made and vibratory effects of the aura were created and changed, so that in-tunement may be made with this vibratory teacher.

Through the times and the years, man's knowledges have been changed to a personal God, which is not correct. Not as a God as you call a God, or term God, the name is not correct. Teacher would be more satisfactory, as stated on other voice tracings. The vibratory effects will need to be corrected, and to be correct so that you may pass through this teacher, learning the lesson as you pass through out into other universal learnings. This vibratory learning for this one thing, is only one small portion of your learning, and it is not a big portion of your learning. It is not the ultimate of your learning. It is only a portion of your learning, to learn how to regulate the frequencies of your bodies. To understand that the aura, or the bodies that have been created to protect and to embody the soul, are only the vibratory effects of the soul. It surrounds, protects, takes care of, and is how this soul is able to change and use frequencies on the third-dimension, to be able to continue its survival there. Is that better understood?

Yes, it is.

As stated before, lessons that are learned in your third-dimensional world are only the aspects that are put into your form that you may learn from. They are of no consequence. What is of consequence is the **tuning of the vibratory fields**. This is all.

When the portal was opened and we came to this dimension, did we leave friends and relations, so they have knowledge of us?

There were no friends and relations as now in your effect are considered friends and relations. There was a closeness there with others, and a vibratory frequency pattern that was set up, but not as on the third-dimension — no, it is not the same. Yes, they do know of the opening and what happened. As told, many did go back and forth, and some did go back and stay, as this was not desired by all. And yes, much of this is now as tales and thoughts, as all do not remember this, and all did not participate in it to begin with, and some are no longer on that vibratory pattern, so it is like old tales. It is like Atlantis now, that is told of and no one is sure. No one there really is sure, because they have nothing tangible.

It's like a myth?

More or less, and if you were to go back into that vibratory frequency it would not be as you were going back to see family or relations, or back to your home. No, this would not be the same, this is totally different.

I can understand why it would be, because birth is only on this third-dimension.

That is not exactly correct, but from your viewpoint I guess it would be.

Then we would only recognize someone we were with on that dimension, through their vibrations, if we were to cross paths?

Now this would be correct, yes.

When two people fall in love, is that because their vibrations match? Or does that have anything to do with it?

That is part, of course.

Are you able to live with a person, and be happy, if your vibrations are really different?

Of course not. How could that be? Vibration is all, as stated many times. You cannot live with someone and your vibrations be completely against each other, unless you change those vibrations. This would be the only way that you could be happy, from your viewpoint.

Then one or the other would have to raise or lower their vibrations to match the other?

That is correct. Thought patterns would also need to be changed.

So lowering wouldn't be a good idea. You would be better off looking for a mate that would raise your vibrations or you could raise theirs to your vibrations?

Yes, that is correct.

But, that is hard to know sometimes.

It is, but it is important to know if they will match, or eventually it will not work.

Like Linda and I found each other out of millions of people? It took some doing, but we did it.

With help. This was for purpose.

Getting back to our questions. If there were people here that would work more with them, would the space people be willing to work with Earth people?

No, they do not desire this.

They wouldn't?

This would not be different, they are not there to interfere. If they go there to become part of you, and change your world, then this would be interference.

I imagine that at this time they are taking samples of water, cattle and many things to see what we are doing to poison ourselves, and to see what will happen to us 500 years from now.

That is basically correct.

I know that during Atlantis we actually worked with them in those days.

That is correct, but you have to understand that things were much different than they are now.

Yes, I know they were.

They were not within the general mass, and the general public eye. This was all done without the knowledge of the public.

Yes, we know that. Can you tell me about Easter Island? You know they have a lot of statues there and no one knows what they were used for.

This was for several reasons. At the time that these were made, this was done, as most things were done then, for the eye and for decorative purposes. It was considered, if something was done, it should be for the eye also. Is that understood?

For beauty?

Yes, for beauty. Also these were to be used — to be within tune — for guidance — would perhaps be what you would understand. For guidance, and for use of more of what you would call electronics or guidance of craft, and also for communications.

Do they emit any kind of energy?

This was to be as receiver and as giver. There are no parts as you call them, like radio parts or electrical parts of any kind within, but these were to be used as tuners could be used. "Others" were to use them, and they were used some.

Was it the type of rock that was used?

That was part.

Can you tell me where the rock came from? Or how they got it?

The rock came from there, but they were not all to remain there. This was not completed. They were to be distributed, to be used separately and also simultaneously. But, this was not completed.

Any reason why it was not completed?

Things changed, their plans were changed. The "Others" decided not to use this.

Where did Edgar Cayce's teachings come from? Did his knowledge come from this dimension totally and was it through guides?

That is correct, yes.

I thought it was. Then it was set up before he returned here?

Something that was desired, yes. Had he not the experiences that he had with working in the temples, with the bodies, and with the experiments that were found in Atlantis at that time, and had he not had the experiences, knowledges and connections from other lifetimes also, this may not have been possible. This would not have been possible for just an average soul. You see, you assume he was an average soul, because he was an average type man. He was a very plain type man in the lifetime that this came through. But, that has nothing to do with the progression of the soul, or with the past knowledges and understandings, as I have so often explained to you.

I think he was in Atlanteah about the same time that "This One" and I were there.

This is very perceptive of you.

Linda and I will research this further through our past lives. I wanted to ask you about the raising of the vibrations of the body here on Earth. Will that make many changes in us? Are we going to get larger? Will our brains get larger?

Many changes are taking place in your body that are not due to the changing of the Earth's vibrations. Many changes are taking place through the nourishment, or lack of nourishment, of the body. This is known. There are changes that will come about in the body that are now being started. As vibratory patterns will be changing for your Earth, they also will affect your vibratory bodies which will in turn then affect the physical body. The changes upon seeing and hearing will be specifically known. As this continues, foods that are now accepted will not be tolerated by the physical body.

Our technology in that field will have to come up tremendously.

Some impurities will not be tolerated. This is already being seen, this is already beginning.

I notice that our boys and girls are getting bigger.

Extremes are more, there are more extremes, both ways.

Then the techology that is going on right now, is something that is really going to be needed a little further down the line?

This will be needed severely.

The doctors now are working on genetic engineering. We feel that a lot of those doctors are reincarnating at this time from Atlantis, and through that era, when they knew about and worked with genetics.

Not only from then, but those from the beginning. Many are coming through now, for the technology is more where they may be of use.

Then genetic engineering should be learned and should be practiced on the Earth?

If it should or should not be, is not for me to say. But, I say this: This is how you were created. This is how you created yourselves. This is how this was done. It has always been done in one way or another. You are doing it to the plants and the animals — there is no difference. Why are you wondering if this is correct for you?

There are a lot of people that are fighting it, mostly the religious people are putting pressure on our politicians to pass laws that they can't do genetic engineering. They say that God is the only one that has a right to work with the body, not us.

Very interesting. Which God do they speak of?

The God they think created us.

There is no such God.

They think that the doctors are playing God.

But they were the ones who created. In a sense, you created yourselves. You see, in the beginning, this was not a question of "Is this right or is this wrong?" The question was "How do we do this best?" There were things that were done that were not successful, but it went on and did not stop. There were those who were not happy, as there are those

181

who are not happy now. There has always been this, where there has been free thinking and free will, this is not a thing that has ever stopped.

Whether you realize it or not, there have always been, either through animals, plants or some way, some type of changes in the human body going on. Also, changes have been made within the vibratory system and the brain and everything that you wish to think of. Changes have constantly gone on, and are going on now involuntarily, through the process of the chemicals that are being inserted into the physical form, that you are told will not hurt you.

People that are there in a hundred years or five thousand years will not be the same as today. This will not be, this cannot be, as nothing stays the same, everything changes. If it is for good or for bad is of no consequence, it still changes.

Yes, I truly understand that. I hope that when I come back things will be changed.

I guarantee things will change when you come back.

When we were talking about the astral body, and the cord that attaches to it, we also talked about the fact that we create most everything for our physical body, also.

You create for yourself, yes. With your energy patterns, vibrations and with your mind.

Good or bad, we create most things for ourselves?

That is correct. You are creating for yourself, so that what you desire may possibly come about. Whether you understand, or realize what you desire, or not, that is of no importance. Your soul knows what you desire, and therefore is trying to create within the third-dimensional world what it is that you desire, so that you may learn to adjust your vibrations. The only reason that this is done is so you may learn lessons, which only means, that you may be guided to do the right thing. To make the right vibratory changes.

I was wondering about after we die — do we have the choice of what we want to do? Can I continue on with my learning there?

This is not easily understood and is misunderstood also. What you consider important, or what you want now, has totally nothing to do with your soul development. What you wish now and what you think of now, is of the third-dimensional plane that you are on — this is only learning for vibratory effects. **Do try to understand this.** This is of more thinking than is normally thought of there, so this may take a little understanding on your part.

The soul development, when not within the third-dimensional effect that you are in, is totally free from the desire and thoughts that you now have. While it is on the other side, it is thinking and looking at what is needed for it in a totally different way. It does not even consider the effects of the Earth at all, or of any third-dimensional thinking at that point. It is interested in its own progression — not with the progression of the body. The body is only what is taken for transportation and learning upon your third-dimensional world — the third-dimensional aspect of the planet Earth.

The true "you" is not who you are now, or what you represent. This is manifestation of what you are doing or learning on that Earth plane now. So, do not think that when you are there, that you will be planning and concerned with what you are now, and what you will be next. No, these are not of concern there, except when you come back into the body to take on another Earth incarnation. Then the decision and the thoughts are there, because then you will have to come back within the third-dimension, and you will need to see what will be needed on the third-dimension to continue your progress.

So, what you were before may not be of any consideration, while again it may be. It may be of great consideration to you then. You may wish to continue on those lines. But, on the true soul development source these things are of no importance, because they do not have a bearing — except how to learn to change the vibratory effect. The planning of what you will do next is only planning in your mind now — is only of your Earthly desires for this time period, which is fine.

It is good to think what you would like to do next and what lines you would go down. But do not think this is the only consideration — that you will create and make what will happen to you next. It may help, and have effect within your vibratory patterns, understand that; but will not be made from your Earthly viewpoint, but from your soul development viewpoint, which is not the same.

Well, I know in the past I had always worked with my mind and inventing things.

This progression has been a progression with your lives, I will tell you that. This has been the path which your soul development has chosen to continue to learn the lessons that it is learning. Through working down the fields that you speak of, that you have feelings and understandings that are now you, these are the ways that have been used to express yourself over and over. Perhaps because they have not totally been learned in effect at this point, or perhaps because they are the quickest way of progression at this point for this soul development. This depends upon the individual soul.

Well, I guess I have to wait until I die to know that.

That is correct, and at that point this will not necessarily be the thing that is considered. This is not only from life to life, but is viewed as an overall picture of what progress can be made through the lifetimes — if certain courses of action are taken, if certain lessons are learned each time. Remember, I have told you how important it is to learn as much as possible each lifetime, and to do the best that you can, as future advancements will be based on what is learned each time.

When talking about the crystal in Atlantis, you mentioned the Great Stone. I have never heard of that. I would like to know what the Great Stone is.

The Great Stone is much as what we have discussed about on Easter Island.

Oh, so it had certain properties in Atlantis also, for communication and that type of thing?

This was used for different purposes.

Has that stone lost its power, at this time?

This is not for now, this has been promised for later when you have better understanding and, until then, I will not tell you because you could not understand, except in very general terms, and that would not be correct. But, this was used much as the Great Crystal and for power sources. There was not only the one, as has been said before. Different things were used for different purposes.

I think it's time to stop for this session.

Chapter Ten

(Tape 20)

When working with Linda, using hypnosis to see auras, you said, "You are changing the mind from which you are observing."

That is correct. When in a normal state of mind you look through your eyes and you are seeing with the conscious — what is given for understanding to the conscious mind. But when you have altered the state of the mind into another consciousness, then you see not through the physical eyes but through the aura. The mind of the aura and the new consciousness, as stated before — not through the third eye as stated on Earth.

I read where some people had made contact with space people, and the space people had said that our technology has exceeded our wisdom and we are going to destroy ourselves.

If you looked within this, this might be within truth. This is not that far incorrect. That is why I am now giving these teachings to you, so that people may be prepared, and their knowledge brought up, so that the Earth's technology and wisdom may come together. This is needed — it has been stated that these two things must advance together.

You said that the space people came from a thrid-dimensional frame. What did frame mean?

This means as a time period. Such as something from another dimension would be, something that would be set aside, that would be different. Like saying, a third-dimensional type of place.

They still had bodies, the same as we have bodies?

They still had bodies, yes.

And when they came into our atmosphere, from the very beginning, they came in with solid forms, bodies such as we know?

That is correct. They all had some type of sheath covering, some kind of body type. They did not all have eyes, as you have eyes, but they did have more perception, and more understanding about the third-dimension than those that they were helping.

One other thing. They always seem to put them as all looking alike. Is that true?

No, that is not correct.

I have a question here I want to ask — well, I think I'll ask something else first —

You do not want the question you were just thinking of?

Which one was that?

About La Mureah, on the way of jails and things of that type.

Oh yes, I want that. I was just going to ask something else first.

In La Mureah, in the early period, the intelligence and the understanding of the people was such, that jails were not needed. For there was no understanding of doing harm to another, or to do deeds that were against one another. In some of the other divisions, of those that were sent to other places, there were problems of this type, and had to be handled for that time period. But, in La Mureah there was no such thing as this. In the late periods of La Mureah, and into Atlantis, there were those who did begin to consider this, as the energies did change and customs did change. Of wanting other's properties, or for something that was not for the good of all — the good of all concerned. There were, at that time, those who still had knowledges and understood the workings of the mind, and could see and knew and understood what other people thought and were doing. The people whose thoughts went against others were then corrected through mind and vibration — through their brain patterns, might be a way of stating this correctly. However, these people were not controlled in any way. They were not dominated nor controlled. This was just as correcting a faulty mechanism so that all may live happily and peacefully. Still, all had their own minds and were within their own thinking. These people were not controlled in any way. That must be understood correctly. The period of Atlanteah, and also of Atlantis, was when much of this, being able to control brain patterns, was lost. As many of those people who had the knowledge did not continue, and many of their understandings were not taught and passed on, and therefore much was lost. But many tales and stories of wizards and sorcerers did come from the thinking of this time, but was much distorted also.

Then they began to use other ways of controlling the people?

That is correct, if control is the correct word. It was not actually control, it was just adjusting a negative wavelength that had been misdirected, into the correct wavelength.

But when we were in Egypt, didn't we still understand that it was just a negative thought that was in the person?

That is true, but the people were not able to make the correction or to even truly understand what happened, therefore, many changes were made. Punishment was then to be used. This had already begun before then, in different places — this was not new. This would not have been a way of Atlantis, but it was already the way in many other places.

189

Then I could safely say that punishment of a negative act will not solve negativity?

It only creates hatred and contempt.

So, what we are doing now by putting people in jail is not really solving the problem?

However, do understand this. You cannot allow those who are doing harm to others indiscriminately to continue. While there should be a free place and a free way for all, it is not right to allow those who do harm to others, that will continue to do this harm, to just go free and continue to do harm. That is not correct either. This becomes a serious problem for you in your dimension.

I know our jails are getting bigger and the courts are more crowded.

The basis of psychology and psychiatry is within a good thinking process, and is a very tiny process that has been started there. At the rate that it is being developed and accomplished, it will not develop sufficiently to make any large amount of progress upon your planet. The time element, and the way it is developed, is going in such a way that it is not going to be of good use. Perhaps of some individual use, yes, but of use in mass, no. This will not be accomplished unless many changes are made and much thinking is altered drastically.

The new knowledges that we are bringing through, will this help any?

This will help, if it is so desired there.

It seems that our answer to things, is to make more problems.

This is because, as spoken before, the wisdom there is not great. This wisdom must be increased. The seeking patterns must be increased, and also changed, before wisdom may be accomplished to be of help there.

190

During World War II, General Patton said he knew he had come back to be part of that war — that it was his destiny. If someone had stopped that war at the beginning, then how would he have been able to fulfill his destiny that he had been working on through his lifetimes?

If the war had not been there, and his particular expertise that he was building not needed, then he would not have been there, or he would have accepted something else for that time. That is all. What he was building, you see, was not only to fight the war, but the knowledge of how to do this in a positive way to bring about what was best for all concerned. What he did, and what he was learning, and was using, was not of negativity. This was not done so that he may get in there and kill people, or so that he might come back and do harm to others. This was to help others to have better understandings, and to save lives and to do what had to be done for the best of all. To help the most people, and how it was to be done correctly — this is what, this is not negative.

If Patton had been building up to this particular war, say in his last three or four lifetimes, would there be karma if someone changed or ended the war right away?

This would depend individually, but if, as I said, the war was not there, then there would be nothing for him to do in that area. He would have to wait to use his particular expertise when there was a time for him to be needed for that, or use it perhaps in some other constructive way.

If someone was working for a particular thing all the time, for several lifetimes, and that thing was about to come about and someone who had no knowledge brought it in, would that be sort of stealing? Would this be permitted?

Interference would not be permitted. This would not be on your dimensional level though — this is speaking on the soul development level.

When you spoke of "When a soul is so developed into one aspect, and there is some small portion that is left, that it may go on and learn that lesson elsewhere." — I took that you meant that all the vibrations were correct, except one vibration, and this they could get in another area, but the soul could go on.

Yes, this is possible. As I have said, there are learnings on the other side, but it is much slower.

Is it possible for a person to do one thing so well, that he could perfect himself in one lifetime?

In one particular lifetime it may appear that only one thing was done, such as music, sports or meditation, but this would not be correct. While this may be the one thing that would be needed to finish, then it may look like only one thing to you. To do one thing well physically does not mean perfection of vibrations for the rest of the soul. Concentration, attunement and dedication to one thing may help to bring into final realization, that would be possible. It would not be possible for a soul to only do one physical thing well and perfect its soul. No, this is not correct.

The time we are on Earth we are more concerned with the vibrations of the body, and after death we are more concerned with the soul, is that correct?

While on your third-dimensional plane and in your body, it is, of course, a fact that you will be of much concern with the body, since you have not had the proper understanding of the soul development. The body has been thought of and considered as the total man. The spirit or the indwelling part of you, and the part that is connected with the portion that is thought of as God (the aura), and the body are all one which makes the human.

The vibrations of the body have only been thought of by you. When not in the third-dimension, or after death, when the soul is stripped of the other bodies and is there, of course the attunement and the thought is for only itself there. It is not the body, for the body does not exist there.

Development of the soul is beginning to come into focus more. Understandings of the soul, and of the body, and of the separation of the two are beginning to come about more. This is what is desired so that you may understand about the two, and balance them more correctly.

What you are saying now is that our vibrations are being brought up, and with what we are learning now, we will be able to help balance our vibrations and the soul on Earth?

Correct, that is what you are trying to do. That is what you are trying to help the soul to develop. Your body is for protection and transportation and learning and understanding on your third-dimensional Earth. It would be as you inside of a car, if the only thing that was considered was the appearance, or the outside of that car, and you inside were never given any concern, then this would not be good. Both the car and the person must be considered. Needs and help and understanding to make both perfect, is what is desired. So here you have not only the soul, but you have the body to help also.

When we experience feelings here on Earth, does the soul experience these feelings, or is it just the vibrations? We know that feelings are turned into energies.

Feelings and emotions were brought into your dimension.

When we came through the portal?

Feelings and emotions are part of the soul also. The soul is not something that is without understanding, without feelings, emotion and so forth. The body also has its different types of feelings, emotions and senses to help you understand and to elaborate on those feelings and emotions. This is all connected and tied in together. Feelings and emotions are part of the soul development. This is why these feelings and emotions must be brought into balance. This is part of the balance. This is why you meditate, and why you learn of negativity and of negative emotion, and positive emotion. So that you may balance them and make them harmonize and handle them correctly within the body, so that you may handle them correctly within the soul. This is a very basic understanding for you now.

Then, when meditating, you are learning to bring all the bodies into balance?

Balancing, correcting, calming of the mind and learning to control the conscious mind, so that there is the correct balance.

Our sub-conscious mind goes on after we die. Is there any part of the conscious mind that goes someplace else?

This all becomes, and is now also, intermingled and one, while still yet separate, as will be explained later. Already I have told of the three minds, and of the intermingling of the minds, bodies and emotions. This is all intermingled to make the one soul development. When this is separated, there is some separation that will be of the soul and basic knowledge.

Conscious mind knowledge does stay in with the astral shell to be dissolved with it. However, the conscious mind knowledge is now also incorporated with, what would be considered, the sub-conscious or super-conscious to where it is now understanding the total evolvement. But, basic conscious everyday things that you do for yourselves — how to drink water, how to get this or that, how to drive a car — these things are kept mostly within the astral shell until they are disintegrated and gone. This is not something that is needed to be brought back again for those exact remembrances in a conscious state — this was not desired. This was done to protect you; remember, I told you about it being done in the beginning? This was made for you and for your protection. While the understandings of all this are now within the sub-conscious, it is not necessary for this to be in the conscious where it is foremost in your mind in another lifetime.

If that were true, you would have all these things within your conscious mind now, of how to chop a tree, how to do this and that. Your conscious mind would be so cluttered with the details of things that you have learned from life to life, that would be of no importance to you now, that the learning of new knowledges would be impossible. While that is still there in the sub-conscious, and if it is brought forward properly, can be remembered, it is not necessary for you to call on it now each and every day.

Is there a certain amount of knowledge bred into us through the genes? The part from the animal that we keep, what we might call self preservation?

That is correct, yes.

I guess it amounts to one thing — breeding does play a big part in our bodies, and if you breed two good ones together, there is a good chance that you are going to get another good one.

Of course, this is known. A good body and a good mind for a soul is of good thinking.

It's known, but no one wants to come out and say it about people.

It is practiced there much with animals and plants, but not with humans.

I would like to ask a couple of questions about clones. First I want to know if it is possible.

Is what possible?

Cloning would mean to take one cell from our body and make an absolute duplicate in all ways.

It has been proven that this would be possible.

If this happened and a human clone were produced, would it have any kind of energy — a soul — or would it be animalistic?

You ask me? Ask yourself! Do you remember the teachings of in the beginning?

Yes.

And now you are talking of making other bodies?

Did we clone them?

195

What would be the difference? You are still creating a body for this to be, for this soul experience to be worked through. This is what was done, not in that exact way as just stated, cloning, no, this was not done in that way then, but still the body was created and arranged and rearranged for best use. Now you speak of creating a body. Would there be a reason why, if a normal functioning animalistic body were produced, that it would not be a body for a soul to experience through?

Yes, there would be one reason, to take parts off, for the original body.

This is not what is stated though. Your desire for what this is to be used for is not the same. This has no meaning within what I just said. The fact that the body is produced, the body is there. You produce bodies now that are not good and still a soul will experience within that body what it can, when it can, for its own purpose. Is that not correct? That has been stated.

Then they really shouldn't be cloned for some of the reasons they talk about, like using it as a doner for a heart, kidneys, etc.?

But understand this. You speak of cloning an identical person. You speak of cloning an identical physical body. A person, the essence of a person is not the physical body. The essence of a person is the soul.

This is true.

You could have a thousand bodies that were exactly alike, you will not have one soul that will divide itself up into a thousand bodies, because the bodies are alike. Different souls would take each body. **THE** body is **A** body, whether it is an exact duplicate or not would not matter.

If the bodies were used for parts, and a soul took it, then would it be like killing a person?

This would be correct.

Then this should not be done for that.

This is a decision I could not make for you.

If this was going to be done, it should be because someone had an extraordinary body and mind and it would be for a soul's development. But we should remember that, if a body is cloned, a soul is going to take it.

This knowledge is given to your Earth. The decision is for the Earth to make.

It seems that, when a person is trying to meditate, the conscious mind likes to stay in control.

It likes to stay in control. It does not wish to give over control to another.

Is this more the animalistic part?

That is correct. This was given for control, for as stated before, this is for everyday duties, taking care of the body, taking care of what is going on exactly now, the more crude portion. The finer portion, the soul development portion is in control of its own things, but not the same as the conscious mind is, with you and your body and your thoughts. It is a different process. While the conscious mind is in control, it does not wish to relinquish control, as one would not want to relinquish control to another country, for a short period even. This is your territory and you wish to make everyone do what you wish, and to still be the controller.

Then what is the best way to convince the conscious mind that you are taking over the control?

As has been stated before, discipline is one of the best ways. Any discipline — meditation, concentration, contemplation — any kind of physical thing that you wish to do to control your body, to control your thoughts. For **you** to control it and not your conscious mind to control it — discipline.

Say two people form a negative karma in this lifetime. Would they have to live this karma out with one another, or can they live with another person that has the same energy, and make the correction in that way? We know karma is energy.

This is preferable, and the most perfect karma is worked out, if it can be between you and that soul development that you are having the problem with. That way it may be resolved and alleviated 100% and completely corrected from both sides at once. That way it does make the most perfect correction. If you do try to do this, and it does not work — say you try to have another life together and you do all you can to correct this, and the other one does not do all they can to correct this, then you have done your portion that you can do. Therefore, the other one's karma will still be there, but your karma may be gone. Is that understood so far?

Yes, that is understood.

Then the other one is still left with their karma for you, while yours may be broken with them. If this is not presented again, then they may be given the opportunity in other ways to work this out for themselves, which will be much more difficult. Still there will be someplace, something, sometime where this new knowledge will be presented to you, as a soul entity, for null and voiding whatever is left to be null and voided. You may not be together to work this out, it may be something that they may have worked most of it out, but still the thread of bond is there, and they may do some good act for you just as a passerby or as something of this type, to give back that good energy and to null that out. This becomes a very intricate thing, but basically, that would be the way it would be handled.

Is karma just in the third-dimension here, or is this something that came through the portal?

No, that is not correct. This is something that, while we speak of a third-dimensional world, of how it is done there, this concept of karma, while not the same as karma on your dimension, the vibratory effects of these things are still carried in other places. As has been stated, entities that are on other dimensions and other levels that come through to work with you now on these levels, do much injustice. While doing the best they can, they do give interference, because even as you know there, they wish to be of importance where they are. This still is within the patterns and the emotions that are there. This is not only a bodily function, but functions of the soul. The vibrations, the emotions and all, are still there, and they may wish to still be important. They produce their own vibratory patterns and effects that must be corrected on that same level. Interference may not be given from any level and go uncorrected. They will have to correct this vibratory effect themselves on the level they are on when the interference is given.

Can accidents on the Earth create karma? Say a person is accidentally killed, does this create karma?

All things are taken into consideration.

Because it may not have been the other person's fault?

This may be a thing that is done for purpose. But, do understand that all things are not for purpose. There still are accidents. While basic laws are followed within the Universe, and basic things are done, each thing is not always done deliberately. There are other avenues and other chances and other aspects. Free will plays a large part in all dimensions.

There is a theory here, that we can create our accidents by bringing together a set of circumstances.

This can be done, and is done often.

Through our own thoughts? In other words, we are giving something energy to happen?

That is true.

Another thing I would like to ask is — after a person dies, if they wish to stay around Earth longer to see what is going on, can they do this for a short time?

If this is so strongly desired, this can be done. This is not normal. This would not be a normal thing to do, but if this was so strongly desired, this could be, yes.

Then it would be of no detriment, if a soul just wanted to see something, and go on?

This would be within the soul, within their development and their desires. There are many that do not have very high vibratory patterns on their soul progression, that do stay and linger within close range of your dimension, within a close dimensional effect.

Are there some that break through from time to time?

This is not desired, but this does happen. It is a fact and is known by you, by your people there.

Yes, I understand. I was just thinking of a soul that had the desire, the curiosity to stay.

It would be possible, but if this was just a curiosity it would probably not be that important. It would have to be strong and for purpose. For you see, there are natural laws and vibratory patterns for everything. This would go against that natural pattern and would be of much detriment. It could be done, you can swim upstream even though the current is very strong against you, if your determination and your body are strong. But it is much easier to flow downstream with the current.

When we were talking the other day, you mentioned something I would like to know more about. You mentioned that, when souls are alone in other dimensions, they are with many others. Are they creating their own world at that time?

This is understood. But this would depend upon the dimension, upon the place that they would be in, and the circumstances again. There are places that you are isolated from the others, although they may be there physically. You are totally shut off, and totally not able to communicate through many different things, many different reasons. One would be to learn the understanding of a discipline.

Even within your world many are with many, and are shut off. Those who have no hearing or do not see or do not speak or ones who have all of those things. They are with many but are totally shut off from the others. There are those that are with others, but their mind is shut off, so there are many different ways even within your dimension that this term could be used. This would completely depend upon the dimension, the world, the circumstance and the learning that is going on there at that time, what that would exactly mean.

In other words, it can be done on almost any dimension, in a sense?

In a sense, yes.

But, that doesn't necessarily mean that people that have come into this dimension blind or without hearing or without these things, choose to take those bodies because this is what they are thinking about?

Mostly they have chosen those bodies for their own soul learning disciplines, yes. Look to the one there that has been so much discussed, the one Helen Keller. This is a very high soul development that chose this thing, for not only a learning process there, but for better understanding for your people. This developed and taught her much, much was gained from this, and much was gained for all on your planet from this also.

This is true. As our vibrations come up, and we learn more about our bodies and perfect them better, and we don't have bodies born blind and crippled, then those particular learnings will not be needed here? Would it just be that at that time people's vibrations will be such that we will not need those learning disciplines?

That can be, and also other ways will be used, other experiences for learning that will be needed.

If a person's guide is giving wrong answers, does this mean that in his last lifetime he was acquainted with the wrong answers, and is not really doing it deliberately?

That would be more correct, yes.

That's what I thought. Then he is taking third-dimensional learning from this dimension into the sub-dimension, and then he is feeding it back to the third-dimension?

That is correct, and that is how much is being distorted there.

Then the Bible was done pretty well the same way?

Much, yes.

Many of the teachings of the Bible came to the people when they were, as we understand now, meditating or deep thinking. So the words given could have been true and could not have been true, depending on what the teachings in the past life were, of the entity that was giving the information?

All these are ways of misunderstandings.

So misunderstandings don't just happen here, they happen on both sides?

Yes. Also, when one talks, another picks out the words that have meaning for them, and the other words are forgotten.

Very true, I'm finding that out. That is why all of your teachings, given to us, will be taken directly from your voice tracings and we will not interject our thoughts.

That will be the best way, for all concerned.

This is all of my questions for today. Thank you.

Chapter Eleven

(Tape 21)

This third-dimension is a vibratory effect isn't it? All things are vibratory effects?

That is correct.

It depends at what rate they are vibrating?

This thing that is called energy or vibration is manifested in your third-dimensional world in many ways. Light, sound, color, solid objects, air, water — each thing is this one thing.

Can we draw energies from one another on this dimension?

Yes, you draw energies from one another.

Can you explain that to me?

It is the way I explained it before, of your vibratory fields that have the pattern that makes you. Within that pattern, if you are a type of energy that burns quickly, and consumes much, and there is need for much more to be consumed, it is possible for you to consume that of others. To draw, to drain, it is all within energy transference for this to be done.

Can you do this without their knowledge?

Yes.

Is this a good practice?

Of course not.

Can you do it without knowing that you are doing it?

This is transference and often is given. This is not a thing of the conscious understanding, this is a thing that is controlled, and is known more by the sub-conscious mind and the super-conscious understanding. This is not done by the conscious mind.

Is there a way of keeping someone from drawing your energies?

Consciously blocking this?

Yes.

This can be done some, yes. But, this is not a thing that is done through the conscious. This is from another level, although interference can be given if set up and incorporated through the mind and the sub-conscious mind into creation. You can give interference there.

I know sometimes I can be around certain people, and I actually feel like they are drawing everything out of me.

That is correct, and this is done often.

The only way I've found to stop it is to shut my mind off from theirs, so I'm not listening to them. Is it through listening?

All energies are given from the body, even through the eyes, is energy given. Attention that is given is also giving energies. Even looking at something is giving your energy to that particular thing. So thinking or concentrating or looking — all of these things remove your energy also.

In other words when you are talking to me, you are actually giving me energy?

204

That is correct.

I've thought this all along, because Linda and I have talked about it. That listening to the tapes seems to build up energy.

Also, you are giving your energy for the attention.

We are exchanging energies in a sense?

In a sense, that is correct.

When we were talking about astral travel, you said, "When you are in a sub-dimension all things change, including time." Is the sub-dimension time slower or faster?

Think of how dreams work. Also, different dimensions have different things on them. Some dimensions and places, you can do either. Time is of a constant thing in your dimension. Time is not always constant other places, although time does exist, it is not always the same.

When I started to compare the time difference between the dream dimension and our dimension, I couldn't come up with a logical answer.

This is not quite understood by you yet — as we discussed going from New York to London in a few moments of your time period, that it would take longer in the other dimension. It could take only a few moments in the other dimension also. But it is also possible to be slowed down to take longer, this is within the dimension, and the division of what you are in, and within the mental body that you are in also. These things cannot be separated easily, because third-dimension, and other dimensions are so different. As you learn about other things, then more dimensional thinking can be put within your grasp and then you will understand it more quickly.

I understand when people astral travel, the minute they realize they are outside their body, they immediately return. Is there something we can do to prevent this?

Practice, as stated before when telling you that "This One"

was redeveloping her sub-conscious mind for seeing auras, because this has not been practiced and used, it needs to be redeveloped. As a set of muscles would need to be developed for specific use. These things have been badly neglected for many incarnations. Now it is time to get the knowledge back, and this may not be something that will come quickly. Sometimes this has been done in past lives, and things may come easier than for one who has never done anything in a past life. The next time it will be much easier, if you find that these things are to be developed more, because you have laid some basic foundations down now. It will be brought forward much quicker the next time. As children you must learn, train and prepare yourselves, and so it is with this. You may not bear the fruits that you desire this lifetime, but much advancement will be made for the next one.

I think I understand all that you have given us. I understand about in the beginning, fully now.

You understand about in the beginning, now?

I think so, pretty well.

Good, then there is more I can tell you.

I would like to hear it.

This is of the beginning, of understanding that the other entities were looking through the window or portal, do you understand?

Looking through the portal, and from one dimension to the other?

Yes, and what they saw in your dimension that you are now on, and then wanting to come through. This you can see within your mind, and understand what it was that they saw to make them want to go there? Is that correct?

Right, they saw the pretty flowers, and animals, etc.

But do understand this — these entities did not have eyes!

They did not see as you see, as you speak of seeing, they did not have form as you have. So, to see through the portal was not to see your Earth as you think of as seeing your Earth. The knowledges that were gained through there were more from sensor type — would perhaps be more of a word so that you could understand — emotional sensors, etc., but not with eyes, as you understand.

Would it be like putting energies into thought? The energies that they perceived around the Earth?

Yes, but they did not see. They did not see. So now you understand that not seeing, and not hearing — that those types of senses were not utilized. So coming into your Earth's atmosphere was not exactly for the reason that you have thought of before. You see, it was not as they looked over and saw these beautiful flowers and things and wanted to come into it. That part of the understanding was not there. Now, when these entities began to work with the animals for what was desired, with remembrances, it was not exactly because they could see and understand what was going on in this new dimension.

This was not totally understood at all, as it was said, the differences of male and female were not totally understood at that time. Not only were not the feelings understood, but the rest of the understanding was not there either. These were some of the problems that were created. When the third-dimensional guides (space people) were sent in, they were needed because these entities were not able to cope and were not able to utilize their abilities there. And, there understandings were not total, because their sensors were not complete for this new dimension. Soul developments that came in could communicate with many other soul developments that were on a vibratory pattern that could communicate with them freely. But the "Others" that were sent there were able to communicate with the entities that came in, and were also of third-dimensional worlds and were able to guide them. Is that more understood now? So, it was a working of the two together, even more, you see, now than was thought before. Total communication between the two, to bring about what was going to be needed there, visible and invisible at this point.

Then the entities took the animal bodies that were prepared for them, and that they had helped to prepare from their level, of course adjustments and readjustments had to be made. But, now try to think of an entity that went into the body, that had been created for it, and all of a sudden he could see! He could hear! The noise was not able to be handled, it was so deafening, it was so horrendous to him. All of a sudden things were brought into view that had never been seen, had never been sensed before. Vision, pictures, things, sights, feelings, smells all began to come in. Total disorientation was there, and still no way to communicate back with those who were not into the bodies yet, and also there was no communication with those who were in the bodies already. There was no language or way to communicate properly.

Total disorientation was there, complete confusion was there. This was a time of great stress and great confusion. No communication could be given to those that were coming in to tell them of these things. Those that were in the body now did not know how to handle these things. Much learning and much horror went on in this period. This was of a very long, hard period in through there. This is why there were many different degrees of understandings. Those that could handle it some, had some understanding and those who completely could not handle this, lost touch of what was before.

Were those some that were damaged and sent back?

Those that went first, that did not wait until they thought they would be able to handle it properly, were severely damaged.

Those are the ones that were sent back for redevelopment?

That is correct, they were very severely damaged. Others later were not as damaged as they were damaged. This was different as things were readjusted and changed by then. What happened was totally without understanding, from anyone's point of view. There were those that had no success, and some that had much success, remember that part? Alright, now do you have a better understanding of what success, and not success, did mean? Not only intelligence, this was not thought of at that point, intelligence was of no thought at that point. To walk into and open your eyes and see and not understand and yet all this was blaring and shouting. Looking, seeing, smelling, feeling, hearing, all this was new, and no verbal or other type of communications. This personality that they went into, of the body, was a very strong personality also.

So there would be a lot of conflict between the conscious animal mind, and their understanding?

That is correct. Total confusion and misunderstanding, because what they understood this world to be, was not at all, what it was. Your dimension was not as it had been perceived by them. It was understood as they could understand it, from their dimensional knowledges. But when they changed, and made the new entity, then they saw your world as it truly is, as seen through the eyes of the animal and the way it is in the third-dimension, and they were not prepared for such, they were not prepared at all.

So the ones that were of strong vibratory frequencies, that were successful, helped all that they could, and also the "Others" did help as much as possible during this period, but still, there were many problems. Much confusion, much total horror and misunderstanding during this period. This was totally different than had ever been anticipated or expected by any.

Then that is part of the reason why they could not return? Because by seeing, feeling, hearing, smelling, all five senses changed their vibratory and their energy patterns to a point that there was no way that they could go back to the other dimension, whether the door closed or not?

This is more correct also. It became harder and harder, as stated, to get back in the vibratory frequencies to go back.

So for us, we would have to learn to handle all these feelings, and bring the energies back to where they were, for us to go back to what we were?

Yes, but this would not be possible now. This is why I said that the problems and the things that were considered at this time were beyond your understanding, at that point, you see? Because you did not understand this yet.

That is true. It helped tie everything together.

Correct, but this would not have been understood, had I given it in the beginning. There would have been too many things for you to have understood and remembered all at that one time.

I cannot only understand it, I can almost feel it.

Alright, so now we have this, and also other problems were, that giving birth was not understood. You see, that male, female, sex, and giving birth was not even understood at that time on the third-dimensional level. It was only what was thought by the soul developments when they came in. Now there are new things that are going on and happening. These new entities that had taken over these animalistic bodies only had the animal instincts that were there. They had no knowledge of what to do with these infants that were born to them. There was no caring for, as is thought of today.

To take care of these newly born entities, other than the basic animalistic part that the Mothers had, was not known. And even into La Mureah this was not formed as a family. This was not as a family thing, this was still as individualized. Each person, or each entity, being its own. Except for the animalistic part, of a more family unit type that was within instinct. There was now the feelings of some of both, but not understood why or how. This was still being — although brought together — was still basically individualized.

The "Others" did much, not only to give the genetic changes and treatments that were needed to be done now, but they also began to take care of these infants for teaching, for mental development and learning also. But during this time many brought total knowledge through, but still had to learn how to live on the third-dimensional world.

They maintained it even after —

Yes, some did. Some were very strong personalities themselves. When the basic confusion left, and the disorientation left in the beginning, then the knowledges were retained you see. And even some that were sent away to other places, some realizations came to them later. Not at first, and not for long periods, since lifetimes were very long then, more than what it is now, understand. As confusion left and "Others" helped to work with them, to control their minds and things, some remembrances were brought back and development, in many places, was quite rapid.

For a very long time in La Mureah, the tiny child was not taken care of and kept, and nurtured in the family pattern at all. It was more taken into, as a growth group, as a unit, and cared for and developed. So family patterns and structures were developed slowly there. Also, language had to be created, some form of communication had to be created, and worked on, and taught and made available to them.

Is that why there are so many languages?

A lot, yes. Therefore, these were some of the problems that happened. It wasn't just like people going somewhere new, and setting up camp and start working out new problems. The problems they had were basic, they were still working on the body and mind. This was a very long and difficult period. It was not an easy thing to come into the new dimension and make a new soul development.

At this point, the soul entities that came into Earth, were they like, what you would call, a younger entity? Younger than the ones that helped them? (space people)

No, they were just of a different dimension. The soul is what was coming through, the soul development. Now you also have a body development, with a soul development that makes the one. So, now you have a better understanding about in the beginning?

I can understand the confusion — it must have been scary, but exciting all at the same time.

It was with great trauma and shock. This was not easily done.

Especially some of the bodies they entered and found themselves in, before the agreement was made to only take one body. Some of them we don't even know, because as you said, "They have disappeared from the Earth."

But some of those were easier; your understanding of their understanding is not complete. You look at what you think their understanding is, and what you know of it, and see from your own level, and point of view. But this is not completely true either. What an animal would see and know and understand from his world, is not the same as you would understand.

So at that time, they had the animal's understanding to contend with.

For some, for short periods for some, but this did not prepare them for the one that they wished to develop there. This did not prepare them for what was to come.

> NOTE TO THE READER: From this point on, a lot of the questions that are asked have been asked before. We thought it would be helpful for the reader to review them.

Where did the idea for the changes in the bodies come from?

This came from total remembrances that the soul itself had. Total remembrances of different times and other times, that it had been on places of other developments, that it had used and utilized at other times. Such as it would be on a third-dimensional development. Also, later the "Others" helped.

Some of the other bodies we entered before we came here, were of physical form?

That would be correct.

But they had never experienced entering animals' bodies and seeing it from an animal's point of view?

All was changed when this happened.

I think what I'm trying to bring out is, that we could not see at this particular time.

That is correct, remembrances are within, and understandings are within. But actual experiences change. Even remembrances of things that you can now remember are changed, are nulled by you. You remember it, but it is not as it actually happened. The emotion, and things of that type, are not actually still with that remembrance. As it is told, women forget from birth to birth. After birth is over that experience is not remembered with the intensity. It is remembered that there was pain, but it is not remembered with the intensity. If it were remembered with that intensity, over and over again, there would probably not be as many children born.

Would it be that the energy is still there but it is very weak?

It is the memory. It is not the experience, and much of that is forgotten. And so it is with the soul, you see. The actual things were remembered, but not the actual feelings, since those were not the types of feelings that were within the soul development then. There were no bodily feelings, it was not the same. As you cut yourself — and in a few years you would remember that you cut yourself, and that it hurt, but you would not remember exactly each pain and each throbbing. Do you understand that?

Yes, I can understand that and how they could have forgotten it.

And, since the body experience was not theirs at that time, and had not been for probably very long periods, this was not understood well.

I imagine some of the animal souls had problems too. I guess it bothered their vibrations a lot.

This is what happened, and this is why much was done for the changing of the mutations that were there. The vibratory effect of those entities mingling with them did cause much damage to the genetic structure of the genes.

So, when we came here, there was actually no sex involved, period? Sex would only be a normal thing of what animals did before we came?

This is more to the truth, of course, that is what I said. While it was part of everything, this was not one specific thing that was of concern.

Then what happened is, when the entities entered the animals they couldn't go back and tell the others that there was a problem?

That is correct.

So they just kept entering and the problems just kept getting worse, and no one really knew what it was?

This was only handled through the intelligence of the souls that came through, and through help from some of the "Others".

How was the problem first known? How did it get back that it was a problem?

The problem did not get back, as a problem gets back to be corrected or something of this type. This was as a shock to those that took the full bodies for the first time, and adult full bodies.

Then before the bodies were taken, there was a point where they came in and —

When this was done, it was done.

At one time, you said something about them coming and going, and this would have been before they entered the animal's body, right? Did they come through the portal and then go back sometimes?

Are you speaking of coming and going through the portal or through the body?

Through the portal, not through the body.

This was before they took the body. This was before the body was taken.

Could we cover that a little bit? Did we come down and look around and see what was there?

As best as could be done with sensors — what you call utilized — at that time, of feelings and emotions and of sensory understandings. As much as is done on the other side now.

Then after that was going on for a while, somebody decided to enter the body?

As much has been stated before, this was not totally understood. And while it was not seen with eyes, the way you see now, it was known and understood, as best it could be, with their understandings from their dimension. This is hard to explain, because I cannot explain to you the total understandings or the goings on of the soul, or the total understanding of the other dimensions — other than trying to tell you small parts at a time, so that you will eventually understand. It was as though they could see, but not with eyes — what their sensors told them and described — and they could create a visual image of what they thought things were. But it was not as when they could see with the eyes of the animals.

Then everything emits an energy, and the energy was what they were picking up?

Yes, that would be a fair description.

So when they picked that energy up, they could apply that energy to whatever thought they could understand?

This is only as a crude analogy for you, but as if someone

that was blind could feel with their hands and their fingers and their emotions and their senses and try to picture in their own mind, create what they thought things were. And then, all of a sudden, they were able to see, and it wasn't what they thought it was at all, was nothing as they thought.

I can imagine, but one thing I want to understand. You used the term "see through the portal," but we know now that's not true.

This was to help you at that time to understand what was going on then, and you had enough other things to try to understand then. But it was not with eyes, no, it was with energies, feelings and sensors. And they perceived it with their own thought forms, they perceived within their own thought which was, within a small degree of reason, within some understanding. Just as now while the soul has the total understanding of what was within that dimension, you consciously do not understand it. But the soul, when it is only a soul, and is within its own development, it would understand.

But not on this dimension?

No, because the soul has now been created into a totally new entity.

I imagine that there are just as many new adventures out there as the last one we went through.

That is correct.

It's very interesting. It really ties a lot of thoughts, ideas, and understandings together and will help many people to understand what has happened to us.

Sight is not necessary other places, this is only a thing for you here, now.

Yes, so there is just a tremendous amount of learning.

Variations and learnings, different combinations, different things. Things that you know absolutely nothing of, nor could you perceive it or understand it, even if it was explained to you in great detail. This is why is it very hard to discuss these things.

Now can you see another reason why the Bible has been distorted? When they had to speak of things that had to be put into words that third-dimensional people of those times could understand, it was difficult. The reason that things were not exactly true, and the truth has not correctly been told, is because it was not understood. As there are things that I have not told you yet, because at this point there is no way you could understand them.

Or accept them?

That is correct. So, as you go along and understand better, then more things will be given. This is why many things that you are curious about are not always given. And while I do not like to continue to say, "This will not be given" or "I will tell you later," these things are of good reason and I hope you will begin to understand that I am not withholding knowledge. Your background is being built quickly now and new thoughts, and ideas and total understandings will be given even quicker now.

Don't stop here, the next three pages are important!

TO THE READER

We wish the reader of this book to understand the knowledge that is coming through belongs to the world, and that we are very happy, that in this lifetime, this connection was brought about by us. It is now known that new and correct knowledge is badly needed on Earth. Not that we feel the knowledge given at this time will save the world, but it may give the people of Earth something to build a better foundation for a better existence on the Earth.

At this time we will assume that you have read this book in its entirety. We cannot stress enough the importance of your reading this book more than once. Before it was published we had numerous people read it and all commented, that after they read the manuscript, they understood themselves and life on Earth much better.

At the time this book was published, we had already received, from the Guardian For Our Universe, knowledge that goes far beyond anything that appears in this book. It was very apparent to us that the knowledge in this book was needed to build the foundation to understand the knowledge that will be following in books to come.

Please read the next page. . . .

QUESTIONS MAY BE ASKED
IN PERSON

As the connection remains open and clear

After due consideration, the Guardian will accept others to be present during the asking of questions, and those present may ask questions also. It is preferred that groups would select 4 or 5 persons to represent your group.

QUESTIONS MAY BE ASKED BY MAIL

Questions may be sent by mail. If you have questions in your mind that have been troubling you, and are of a general nature, please ask by mail. We will in turn ask the Guardian, and the questions and answers will be put in books that follow, as we intend to bring knowledge forward that is of interest to all. Please enclose a statement with your question stating if you **do** or **do not** want your name used in a book as the one who asked the question.

COMMENTS AND QUESTIONS
TO THE AUTHORS

As being the authors and bringing this knowledge to you, your comments and questions to us personally will be welcome. Please direct your letters to whom you wish to answer, either Fred, or Linda. If you wish an answer, please enclose a large, self-addressed, stamped envelope.

TAPES

Those who read the manuscript wanted to listen to the original tapes. After listening to the tapes, they felt the tapes contributed to the raising of their vibrations and they assimilated the information much quicker and they thought that we should make them available to the readers. Therefore, the original tapes will be made available if asked for. The original tapes have conversation between the Guardian and Fred that is of a general nature and was not put in the book. We want you to understand this.

BOOKS

When ordering additional copies of **"Guardian 1, The Answers,"** please send $9.95 (plus $1.00 for shipping and handling) for each additional copy. (California residents add 6 percent sales tax.)

SEND ALL INQUIRIES TO:

Fred Foster Publications
5200 Stockton Blvd. #155
Sacramento, CA 95820

Mail to: Fred Foster Publications
5200 Stockton Blvd. #155
Sacramento, CA 95820

☐ Please send me additional information concerning the tapes that are available.

☐ Please send me _____ copies of **"Guardian 1, The Answers."** I am enclosing $9.95 (plus $1.00 for shipping and handling) for each book ordered. (California residents add 6 percent sales tax.)

I am enclosing a total of $_____.

Make all checks payable to: **Fred B. Foster, Publications**

(PLEASE PRINT)

_____
NAME

_____
STREET ADDRESS

_____
CITY   STATE   ZIP

Mail to: Fred Foster Publications
5200 Stockton Blvd. #155
Sacramento, CA 95820

☐ Please send me additional information concerning the tapes that are available.

☐ Please send me _____ copies of **"Guardian 1, The Answers."** I am enclosing $9.95 (plus $1.00 for shipping and handling) for each book ordered. (California residents add 6 percent sales tax.)

I am enclosing a total of $_____.

Make all checks payable to: **Fred B. Foster, Publications**

(PLEASE PRINT)

_____
NAME

_____
STREET ADDRESS

_____
CITY   STATE   ZIP

Meats - 77
UFO's   90
Pryamid   93
Transition-95

Walk-Ins 129
Jesus   130